MOLECULAR VIROLOGY

Edited by **Moses P. Adoga**

INTECHOPEN.COM

Molecular Virology
Edited by Moses P. Adoga

CBS, Edition **2017**

Published by InTech

Janeza Trdine 9, 51000 Rijeka, Croatia

Publishing Process Manager Ana Skalamera

Technical Editor Teodora Smiljanic

Cover Designer InTech Design Team

Additional hard copies can be obtained from orders@intechopen.com

Molecular Virology, Edited by Moses P. Adoga
p. cm.
ISBN 978-953-51-0369-1

INTECH

open science | open minds

Contents

Preface

Since the early experiment conducted by Dmitri Ivanovski in 1892 that demonstrated how filtered sap from diseased tobacco plants had infectious particles that subsequently infected healthy ones and the later naming in 1898 of the infectious particles by Martinus Beijerinck as *contagium vivum fluidum,* from which the name virus derived, the study of viruses and virus-like agents (Virology) has over the years advanced in experimental depths and sophistications. It was this advancement that led to a molecular dimension to virology with much greater significant discoveries being the continual outcome. This book is therefore, intended to serve as a major contribution to this fast advancing field of study.

Chapter one discusses HIV-1 reservoirs and latency. The authors treated it against the backdrop that reservoirs have been reported to serve as major impediments to a cure and at the same time offer useful clues for eradication. Researchers concerned with HIV-1 eradication studies will find this part of the book of utmost importance. In chapter two, the authors did justice to the molecular evolution of hepatitis viruses including their genetic diversities, and further exposed the fact that such evolutionary diversities are an impediment to vaccine initiatives. Users will discover in chapter three the metabolic disorders that can result from hepatitis C virus infection. The authors tried to make the readers understand that although the primary target of hepatitis C virus (HCV) is the liver, metabolic complications have also been associated with its infection. This chapter also did justice to the role of chronic HCV infection in proteinuria, type-2 diabetes, metabolic syndrome and lipid abnormalities among several other aspects.

Chapter four explores the genome of the influenza C virus and discusses the structure and function of its M gene. Coding strategies of influenza C virus M gene, regulation of M1 and CM2 synthesis, and the role of M gene products in virus replication are also discussed inter alia. Viral vectors and their applications in therapeutic research are discussed in chapter five. The authors treated their applications in neurobiology and gene therapy, among others. In chapter six, various aspects of avian influenza are discussed with emphasis on why it is cost effective to focus more efforts on preventive measures against spending much more on treatment (when prevention should have cost less). Chapter seven discusses how Japanese encephalitis virus infection can be prevented with bella-donna 200. The authors also reported their research experience with this product.

Virologists with special interest in insects and all those concerned with agricultural pest studies and control will find the last two chapters of great importance. In chapters eight and nine, the authors discussed how the field dynamics of insect populations can be influenced by viruses that are pathogenic to insects. Treated are the role of baculovirus enhancins in pathogenicity, identification and analysis of the enhancins, their function and how polydnaviruses interact with nucleopolyhedroviruses in Lepidoptera larvae inter alia.

The great contributions in this book by highly experienced scientists including renowned professors are a clear indication that from the first to the last chapters, you are up for a thrilling new insight into Molecular Virology. The publisher, editor and the authors hope that researchers, lecturers and students alike will find this masterpiece an invaluable companion.

Moses P. Adoga
PhD Cellular and Molecular Biology
Bangor University, School of Biological Sciences
UK

1

HIV-1 Reservoirs and Latency: Critical Barriers and Clues for HIV / AIDS Cure

Moses P. Adoga[1,2] and Grace Pennap[1]
[1]*Microbiology Unit, Department of Biological Sciences, Nasarawa State University, Keffi,*
[2]*Department of Bioinformatics, University of Leicester,*
[1]*Nigeria*
[2]*UK*

1. Introduction

Human immunodeficiency virus type 1 (HIV-1) infection is still a formidable threat to public health in spite of remarkable therapeutic advances (Adoga et al., 2009; Goselle, 2011). Highly active antiretroviral therapy (HAART) is effective against HIV/AIDS but it is not curative, and hence does not lead to the eradication of HIV-1 infection. Patients on HAART usually have their HIV RNA levels suppressed below the lower detection limit of 50 copies / ml. However, interruption of treatment leads to viral rebound and HIV can be detected again usually within two weeks (Davey et al.,1999; Dahl et al., 2010). This observation led to the initial questions that spurred investigations into the discovery of the critical roles of reservoirs and latency in HIV-1 persistence.

Scientists continue to demonstrate that the major barriers to HIV/AIDS cure are the latently infected CD4+ T lymphocytes harbouring replication-competent HIV in lymph nodes, spleen and cells of the CNS (Cordelier, 2006; Dinoso et al., 2009). HIV-infected patients harbour ~ 10^5 to 10^6 memory CD4 T-cells that contain fully integrated but transcriptionally silent HIV proviruses, which constitute a reservoir of viruses that show no sensitivity to highly active antiretroviral therapy (HAART), leading to HIV persistence in patients for life (Williams and Greene, 2007).

Latently infected cells may be defined as transcriptionally silent cells that contain integrated HIV DNA, but which are capable of producing infectious virus only upon activation (Lassen et al., 2006; Pace et al., 2011). Based on this definition, it may be deduced that latently infected cells do not transcribe HIV RNA or express HIV proteins. On the contrary, it has been reported that resting CD4+ T cells and PBMC (peripheral blood mononuclear cell) from patients taking HAART do contain low levels of HIV RNA (Li et al., 2005; Fischer et al., 2008; Pasternak et al., 2009). However, the low levels of HIV RNA are lower than in activated CD4+ T cells, which brings to question as to whether the low levels of HIV RNA is enough for significant expression of protein (Lassen et al., 2004a; Zhang et al., 2004).

Current evidence suggests that HAART suppresses viremia to below clinically detectable limit (50 HIV-1 RNA copies/ml) (Dinoso et al., 2009). It has been reported earlier that antiretroviral drugs caused increase in CD4+ T cell counts and exponential decay in viremia, demonstrating the short lifespan of plasma virus with half life of about 2 days (Wei et al., 1995; Perelson et al., 1996).This led to the earlier idea that, with prolonged use of combination antiretroviral drugs, HIV-1 could be eradicated completely leading to a cure (Perelson et al., 1997). However, we now know that mere prolonged use of HAART cannot cure HIV/AIDS because HIV-1 is capable of persisting in latently infected resting $CD4^{+}$ T cells which constitute a reservoir for the virus (Rong and Perelson, 2009; Jochmans et al., 2010; Margolis, 2010).

This leads to the clear challenge of eliminating the virus in reservoirs or sanctuaries before a cure, hence prompting scientists in this field to devise various strategies or models of purging and eliminating the virus from the reservoir. In this chapter, we discussed HIV-1 persistence and latency, the mechanisms for HIV latency, dynamics of persistent viremia, dynamics of viral decay, and current approaches for purging latent HIV reservoirs inter alia.

2. Defining HIV-1 reservoir

A cell type or anatomical site where a replication-competent form of a virus persists for a longer time than in the main pool of actively replicating virus can be termed a viral reservoir. In the case of HIV-1, $CD4^{+}$ T-cells serve as major reservoirs both during HAART and untreated infection. The ultimate goal of HIV therapeutic interventions is to eradicate HIV-1 from persons infected with the virus. The development of potent antiretroviral regimens that greatly suppress HIV-1 replication has witnessed important life-saving advancement.

Despite these therapeutic advances, major obstacles remain to eradicating HIV-1. Reservoirs of HIV-1 have been identified that represent major impediments to eradication. Conceptually, there are 2 types of sanctuaries or reservoirs for HIV-1, cellular and anatomical. Cellular sanctuaries or reservoirs may include latent $CD4^{+}$ T cells containing integrated HIV-1 provirus; macrophages, which may express HIV-1 for prolonged periods; and follicular dendritic cells, which may hold infectious HIV-1 on their surfaces for indeterminate lengths of time. HIV-1 infected patients harbor ~ 10^5 - 10^6 memory CD4+ T cells that contain fully integrated but transcriptionally silent HIV proviruses. The key anatomical reservoir for HIV-1 appears to be the central nervous system. Critical clues for HIV-1 eradication lie in the understanding of the dynamics of HIV-1 within these reservoirs (Schrager & D'Souza, 1998; Williams & Greene, 2007).

2.1 Resting CD4+ T cells are the main contributors to the reservoir

The description of latently infected resting CD4+ T cells was first done by Chun and his colleagues (Chun et al., 1995). Latently infected CD4+ T cells are resting CD4+ T cells that lack activation markers including CD25, HLA-DR and CD69. Resting CD4+ T cells are in the G0/1a stage of the cell cycle, express limited levels of the transcription factors NFAT and NF-κβ, and have limited pools of deoxynucleosides, which are important for the efficient expression of the HIV LTR. Besides, resting CD4+ T cells have been shown to be enriched for microRNAs involved in HIV latency (Han et al., 2007; Huang et al., 2007; Colin & Van Lint, 2009; Margolis, 2010). In addition, studies have shown that during acute infection when reservoir establishes, the most prominently infected cell types are the resting CD4+ T cells; and CD4+ T cells are the most frequently infected before or during HAART. An in vitro study with enhanced HIV-1 integration assay has also reported that HIV-1 integrates into resting CD4+ T cells even at low viral inoculum, suggesting that there is no threshold number of virions required for integration into resting CD4+ T cells (Schacker et al., 2000; Li et al., 2005; Agosto et al., 2007). Figure 1 illustrates dynamic patterns of the infection of CD4+ T cells.

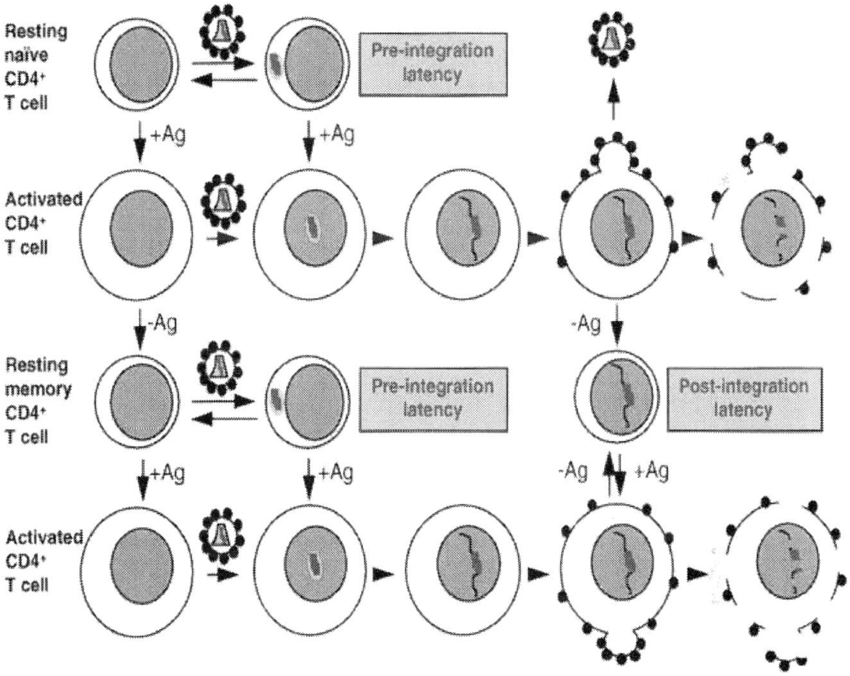

Fig. 1. Dynamics of HIV-I Infection of $CD4^+$ T lymphocytes: The horizontal arrows show the successive steps in the life cycle of the virus; while the vertical arrows show the transitions between resting (small) and activated (large) CD4+ T cells. X4 isolates of HIV-1 can infect resting and activated CD4+ T cells, but replication does not occur in resting cells due to a block prior to the stage of nuclear import of the pre-integration complex containing the reverse-transcribed HIV-1 DNA. R5 isolates can infect activated CD4+ T cells, but may infect only the subset of resting CD4+ T cells that express sufficient amounts of CCR5 (Finzi & Siliciano, 1998).

3. Dynamics of HIV-1 infection, decay characteristics and latency

The dynamics of viral replication in vivo offer the best context to consider the pathophysiology of HIV-1 infection and the mechanisms of viral persistence.

Understanding viral dynamics requires a steady state (in which continuous virus production is balanced by virus clearance) analysis of the amount of free virus and the number of virally infected cells present in infected individuals (the viral load) and a dynamic analysis of the rates at which virus particles and virally infected cells are generated and cleared. Substantial progress has been made in understanding key elements of HIV-1 dynamics, and the perspectives developed have already proven useful in understanding the pathogenesis of other infectious diseases.

Several studies have described the dynamics of HIV-1 infection, decay characteristics and latency. In one of such studies, Perelson and colleagues used a set of differential equations and described the dynamics of cell infection and virion production and by fitting the decay data to the derived model. This way they were able to examine plasma viral levels early after the initiation of therapy in an effort to measure separately the two processes that contribute to the rapid initial decay of the plasma virus: the clearance of free virions and the loss of the infected cells that produce most of the plasma virus. They found out that corresponding half life ($t_{1/2}$) values for free virions ranged from 0.18 to 0.34 days , with a mean of 0.24 ± 0.06 days (~ 6 hours).

Total daily virion production and clearance rates range from 0.4 x 10^9 to 32.1 x 10^9 virions per day, with a mean of 10.3 x 10^9 virions per day released into the extracellular fluid. The average life span of a virion in the extracellular phase is 0.3 ± 0.1 days, while the average life span of a productively infected cell (presumably an activated CD4+ T cell) is 2.2 ± 0.8 days. Productively infected CD4+ T cells are lost with with an average t ½ of about 1.6 days. The average life span of a virion in blood was calculated to be 0.3 days. Therefore a population of plasma virions is cleared with a t ½ of 0.24 days. This implies that, on the average, half of the population of plasma virions turns over in about every 6 hours. The average generation time of HIV-1, defined as the time from the release of a virion until it infects another cell and causes the release of a new generation of viral particles, was determined to be 2.6 days (Perelson et al., 1996; Finzi & Siliciano, 1998; Pierson et al., 2000; Dahl et al., 2010).

4. How HIV-1 reservoirs are formed

Several mechanisms may be involved in the formation of reservoirs. One mechanism is the direct infection of resting CD4+ T cells. This is supported by the fact that resting cells are the prominently infected cells during early infection (Li et al., 2005). Still in support of this mechanism, data from in vitro studies have also demonstrated that it is possible to infect resting cells directly (Agosto et al., 2007; Plesa et al., 2007; Dai et al., 2009).

In addition, a cytokine-rich environment and the presence of macrophages may play some roles in enhancing the formation of reservoir through some mechanisms (Eckstein et al., 2001; Swingler et al., 2003).

Another idea is that latently infected CD4+ T cells may originate from activated CD4+ T cells that return to a resting state after becoming infected. The fact that memory CD4+ T cells contribute the most to latently infected CD4+ T cells lends credence to this idea (Chun et al., 1997; Han et al., 2007).

4.1 How to measure HIV-1 reservoirs

The assay used in measuring latently infected cells was developed by Siliciano and Wong with their colleagues, and is referred to as the Infectious Units Per Million (IUPM) assay.

This assay involves the serial dilution and subsequent activation of resting CD4+ T cells from HIV-infected patients on HAART in the presence of allogeneic susceptible T blasts as targets to allow spreading infection (Finzi et al., 1997; Wong et al., 1997; Siliciano & Siliciano, 2005). Determining the number of latently infected cells is made possible by enumerating the number of wells that demonstrate positive results for spreading infection under limiting dilution conditions. The IUPM assay, though costly and laborious, has proved highly invaluable for the characterisation of reservoir cells. Using this assay, latently infected cells are defined as cells that contain HIV DNA but do not produce infectious virions until they are stimulated to enter the cell cycle. Perhaps a better description is to define latent cells as those cells that contain integrated DNA that do not release virions until stimulated, if steps are taken to remove pre-integration complexes (Lassen et al., 2004a; Lassen et al., 2004b).

Studies have shown that measurement of integrated HIV DNA is a useful surrogate marker of latently infected cells. However, studies that involved measuring latently infected cells over a period of time demonstrated constant level of both IUPM and integrated HIV DNA. This observation suggests that changes in the levels of integrated HIV DNA would be a good surrogate marker for changes in reservoir size (Brussel et al., 2003; Brussel et al., 2005; Koelsch et al., 2008; Chomont et al., 2009; Richman et al, 2009).

The integration assay appears to be more sensitive and easier to perform than the IUPM assay, but the IUPM assay will best determine if any cells are capable of producing viable HIV particles. Some studies have demonstrated the superiority of integration assay over the IUPM assay. For instance, the levels of latently infected cells are so low in elite suppressors that the level is below the detection limit of the IUPM assay. However, a particular study detected integrated DNA in 10 out of 10 elite suppressor individuals (Blankson et al., 2007; Julg et al., 2010; Graf et al., 2011). The fact that a large fraction of integrated HIV proviruses are defective and contain a large number of mutations should be brought into consideration when using an integration assay as a surrogate marker for latently infected cells.

5. How to purge latent HIV-1 reservoirs

Most strategies for purging latent HIV-1 reservoirs involve the activation of latently infected cells to induce the expression of the integrated HIV-1 DNA making it susceptible to antiretroviral therapy and immune-mediated killing. Latently infected cells can be reactivated through a number of ways. One way is to up-regulate cellular transcription to induce HIV-1 gene expression. This involves inhibiting HDACs which promote latency by regulating genome structure and transcriptional activity. A study by Archin and colleagues demonstrated that synthetic HDAC inhibitors are capable of reactivating latently HIV-infected cells in vitro (Archin et al., 2009). However, when HDAC inhibitor, HAART and valproic acid were co-administered, it gave mixed results (Lehrman et al., 2005; Blankson et al., 2007). Secondly, some interleukins have shown some promise in controlling latently infected cells. For instance, Chun and his colleagues demonstrated that patients treated with IL-2 plus HAART had fewer resting memory CD4+ T cells than patients who had HAART alone . The role of IL-7 in purging latently infected cells has also been tested in some studies with significant results. In addition, it is possible to purge latently infected cells by combining DNA methylation inhibitors with HAART, since DNA methylation is known to reduce intracellular transcriptional activity (Chun et al., 1999; Kauder et al., 2009). Prostratin has also been shown to up-regulate HIV-1 expression in peripheral blood mononuclear cells from patients on HAART but down-regulates the expression of CD4 receptor; and another study has demonstrated that the transcription factor *Est1* reactivates latent HIV-1 in resting memory T cells in patients who were on HAART without resulting to general activation of T cells (Kulkosky et al., 2001; Yang et al., 2009).

Another method is to increase HIV-1 gene expression by altering the effects of non-coding cellular miRNAs. There are reports of cellular miRNAs contributing to HIV-1 latency in memory CD4+ T cells. When cellular miRNAs bind to the 3′ end of HIV-1 messenger RNA, they inhibit viral protein translation in cells resulting to the enhancement of latency.

However, combination of the different activators may play a synergistic role in the reactivation of latent viral reservoirs as demonstrated by some studies. For instance, Reuse and colleagues have demonstrated that combining valproic acid and prostratin or suberoylanilide hydroxamic acid (SAHA)- an HDAC inhibitor- and prostratin more efficiently reactivated HIV-1 production in cell lines and cells isolated from patients receiving HAART than each compound alone (Reuse et al., 2009; Dahl et al., 2010).

6. Conclusion

Latent HIV-1 reservoirs are established early during primary infection in lymphocytes and macrophages and constitute a major barrier to eradication even in the presence of highly active antiretroviral therapy (Alcami et al., 2010). HAART reduces HIV-1 in plasma to undetectable levels, which led to the earlier idea that prolonged treatment might eradicate the infection. However, it was later discovered that HIV-1 can persist in a latent form in resting CD4+ T cells. In addition, both cellular and viral miRNAs could be involved in maintaining HIV-1 latency or in controlling low-ongoing viral replication. Identification of new cellular elements restricting the viral cycle provides a new paradigm on HIV-1 latency (Alcami et al., 2010). It is obvious from growing evidence that HIV-1 eradication cannot be achieved without addressing the vicious circle of HIV-1 latency and reservoirs. Latently infected cells serve as a constant source of viral rebound even in the face of antiretroviral therapy. Finally, HIV-1 reservoirs and latency present monumental challenges to the scientific community, especially those involved in therapeutic research; and at the same time offer useful clues for eradication.

7. References

Adoga, M.P., Banwat, E.B., Forbi, J.C., Nimzing, L., Pam, C.R., Gyar, S.D., Agabi, Y.A., Agwale, S.M. (2009). Human immunodeficiency virus, hepatitis B virus and hepatitis C virus: sero-prevalence, co-infection and risk factors among prison inmates in Nasarawa State, Nigeria. *Journal of Infection in Developing Countries*, 3 (7): 539-547.

Agosto, L.M.,Yu, J.J., Dai, J., Kaletsky, R., Monie, D. & O'Doherty, U. (2007). HIV-1 integrates into resting CD4+ T cells even at low inoculums as demonstrated with an improved assay for HIV-1 integration. *Virology*, 368 (1): 60–72.

Alcami, J., Coiras, M., Lopez-Huertas, M.R. & Perez-Olmeda, M. (2010). Molecular mechanisms involved in HIV latency and implications for HIV treatment and eradication. *Retrovirology*, 7 (suppl 1): 115.

Archin, N.M., Keedy, K.S., Espeseth, A., Dang, H., Hazuda, D.J. & Margolis, D.M. (2009). Expression of latent human immunodeficiency type 1 is induced by novel and selective histone deacetylase inhibitors. *AIDS*, 23: 1799-1806.

Blankson, J.N., Bailey, J.R., Thayil, S., Yang, H.C., Lassen, K., Lai, J., Gandhi, S.K., Siliciano, J.D., Williams, T.M. & Siliciano, R.F. (2007). Isolation and characterization of replication-competent human immunodeficiency virus type 1 from a subset of elite suppressors. *Journal of Virology*, 81 (5): 2508–2518.

Brussel, A., Mathez, D., Broche-Pierre, S., Lancar, R., Calvez, T., Sonigo, P. & Leibowitch, J. (2003). Longitudinal monitoring of 2-long terminal repeat circles in peripheral blood mononuclear cells from patients with chronic HIV-1 infection. *AIDS*, 17 (5): 645–652.

Brussel, A., Delelis, O. & Sonigo, P. (2005). Alu-LTR real-time nested PCR assay for quantifying integrated HIV-1 DNA. *Methods in Molecular Biology*, 304: 139–154.

Chomont, N., El-Far, M., Ancuta, P., Trautmann, L., Procopio, F.A., Yassine-Diab, B., Boucher, G., Boulassel M.R., Ghattas G., Brenchley J.M., Schacker T.W., Hill B.J., Douek D.C., Routy J.P., Haddad E.K. & Sekaly R.P. (2009). HIV reservoir size and persistence are driven by T cell survival and homeostatic proliferation. *Nature Medicine*, 15 (8): 893–900.

Chun, T.W., Finzi, D., Margolick, J., Chadwick, K., Schwartz, D. & Siliciano, R.F. (1995). In vivo fate of HIV-1-infected T cells: quantitative analysis of the transition to stable latency. *Nature Medicine*, 1: 1284–1290.

Chun, T.W., Carruth, L., Finzi, D., Shen, X., DiGiuseppe, J.A., Taylor, H., Hermankova, M., Chadwick, K., Margolick, J., Quinn, T.C., Kuo, Y. H., Brookmeyer, R., Zeiger, M.A., Barditch-Crovo, P. & Siliciano, R.F. (1997). Quantification of latent tissue reservoirs and total body viral load in HIV-1 infection. *Nature*, 387: 183–188.

Chun, T.W., Engel, D., Mizell, S.B., Hallahan, C.W., Fischette, M., Park, S., Davey Jr.,R.T., Dybul, M., Kovacs, J.A., Metcalf, J.A., Mican, J.M., Berry, M.M., Corey, L., Lane, H.C. & Fauci, A.S. (1999). Effect of interleukin-2 on the pool of latently infected, resting CD4+ T cells in HIV-1-infected patients receiving highly active anti-retroviral therapy. *Nature Medicine*, 5: 651-655.

Colin, L. & Van Lint, C. (2009). Molecular control of HIV-1 postintegration latency: implications for the development of new therapeutic strategies. *Retrovirology*, 6: 111.

Cordelier, P. & Strayer, D.S. (2006). Using gene delivery to protect HIV-susceptible CNS cells: Inhibiting HIV replication in microglia. *Virus Research*, 118: 87-97.

Dahl, V., Josefsson, L. & Palmer, S. (2010). HIV reservoirs, latency, and reactivation: prospects for eradication. *Antiviral Research*, 85 (1): 286–294.

Dai, J., Agosto, L.M., Baytop, C., Yu, J.J., Pace, M.J., Liszewski, M.K. & O'Doherty, U. (2009). Human immunodeficiency virus integrates directly into naive resting CD4+ T cells but enters naive cells less efficiently than memory cells. *Journal of Virology*, 83 (9): 4528–4537.

Dahl, V., Josefson, L. & Palmer, S. (2010). HIV reservoirs, latency, and reactivation: prospects for eradication. *Antiviral Research*, 85: 286-294.

Davey, R. T., Bhat, N., Yoder, C., Chun, T.W., Metcalf, J.A., Dewar, R., Natarajan, V., Lempicki, R.A., Adelsberger, J.W., Miller, K.D., Kovacs, J.A., Polis, M.A., Walker, R.E., Falloon, J., Masur, H., Gee, D., Baseler, M., Dimitrov, D.S., Fauci, A.S. & Lane,

H.C. (1999).HIV-1 and T cell dynamics after interruption of highly active antiretroviral therapy (HAART) in patients with a history of sustained viral suppression. *Proceedings of the National Academy of Sciences USA*, 96: 15109–15114.

Dinoso, J.B., Rabi, S.A., Blankson, J.N., Gama, L., Mankowski, J.L., Siliciano, R.F., Zink, M.C. & Clements, J.E. (2009). A simian immunodeficiency virus-infected macaque model to study viral reservoirs that persist during highly active antiretroviral therapy. *Journal of Virology*, 83 (18): 9247-9257.

Eckstein, D.A., Penn, M.L., Korin, Y.D., Scripture-Adams, D.D., Zack, J.A., Kreisberg, J.F., Roederer, M., Sherman, M.P., Chin, P.S. & Goldsmith, M.A. (2001). HIV-1 actively replicates in naive CD4(+) T cells residing within human lymphoid tissues. *Immunity*, 15 (4): 671–682.

Finzi, D., Hermankova, M., Pierson, T., Carruth, L.M., Buck, C., Chaisson, R.E., Quinn, T.C., Chadwick, K., Margolick, J., Brookmeyer, R., Gallant, J., Markowitz, M., Ho, D.D., Richman, D.D. & Siliciano, R.F. (1997). Identification of a reservoir for HIV-1 in patients on highly active antiretroviral therapy. *Science*, 278: 1295–1300.

Finzi, D. & Siliciano, R.F. (1998). Viral dynamics in HIV-1 infection. *Cell*, 93: 665-671.

Fischer, M., Joos, B., Niederöst, B., Kaiser, P., Hafner, R., von Wyl, V., Ackermann, M., Weber, R. & Günthard H.F. (2008). Biphasic decay kinetics suggest progressive slowing in turnover of latently HIV-1 infected cells during antiretroviral therapy. *Retrovirology* 5: 107.

Goselle, O.N. (2011). Human immunodeficiency virus transmission, in Venketaraman, V (ed.), *Global View of HIV Infection*, Intech, Croatia, pp 43-66. Available from: http://www.intechopen.com/articles.

Graf, E.H., Mexas, A.M., Yu, J.J., Shaheen, F., Liszewski, M.K., Di Mascio, M., Migueles, S.A., Connors, M. & O'Doherty, U. (2011). Elite Suppressors Harbor Low Levels of Integrated HIV DNA and High Levels of 2-LTR Circular HIV DNA Compared to HIV+ Patients on and off HAART. *PLoS Pathogens*, 7 (3): 10.1371/annotation/0d21de23-d44c-49c0-9a9f-53d421648cbf.

Han, Y., Wind-Rotolo, M., Yang, H.C., Siliciano, J.D. & Siliciano, R.F. (2007). Experimental approaches to the study of HIV-1 latency. *Nature Reviews Microbiology*, 5 (2): 95–106.

Huang, J., Wang, F., Argyris, E., Chen, K., Liang, Z., Tian, H., Huang, W., Squires, K., Verlinghieri, G. & Zhang, H. (2007). Cellular microRNAs contribute to HIV-1 latency in resting primary CD4+ T lymphocytes. *Nature Medicine*, 13 (10): 1241–1247.

Jochmans, D., Anders, M., Keuleers, I., Smeulders, L., Kräusslich, H.G., Kraus, G. & Müller, B. (2010). Selective killing of human immunodeficiency virus infected cells by non-nucleoside reverse transcriptase inhibitor-induced activation of HIV protease. *Retrovirology*, 7: 89. Doi: 10.1186/1742-4690-7-89.

Julg, B., Pereyra, F., Buzon, M.J., Piechocka-Trocha, A., Clark, M.J., Baker, B.M., Lian, J., Miura, T., Martinez-Picado, J., Addo, M.M. & Walker, B.D. (2010). Infrequent recovery of HIV from but robust exogenous infection of activated CD4(+) T cells in HIV elite controllers. *Clinical Infectious Diseases*, 51 (2): 233–238.

Kauder, S.E., Bosque, A., Lindqvist, A., Planelles, V. & Verdin, E. (2009). Epigenetic regulation of HIV-1 latency by cytosine methylation. *PLoS Pathogens*, 5: e1000495.

Koelsch, K.K., Liu, L., Haubrich, R., May, S., Havlir, D., Gunthard, H.F., Ignacio, C.C., Campos-

Soto, P., Little, S.J., Shafer, R., Robbins, G.K., D'Aquila, R.T., Kawano, Y., Young, K., Dao, P., Spina, C.A., Richman, D.D. & Wong, J.K. (2008). Dynamics of total, linear nonintegrated, and integrated HIV-1 DNA in vivo and in vitro. *Journal of Infectious Diseases*, 197 (3): 411–419.

Kulkosky, J., Culnan, D.M., Roman, J., Dornadula, G., Schnell, M., Boyd,M.R. & Pomerantz, R.J. (2001). Prostratin: activation of latent HIV-1 expression suggests a potential inductive adjuvant therapy for HAART. *Blood*, 98: 3006-3015.

Lassen, K., Han, Y., Zhou, Y., Siliciano, J. & Siliciano, R.F. (2004a). The multifactorial nature of HIV-1 latency. *Trends in Molecular Medicine*, 10 (11): 525–531.

Lassen, K.G., Bailey, J.R. & Siliciano, R.F. (2004b). Analysis of human immunodeficiency virus type 1 transcriptional elongation in resting CD4+ T cells in vivo. *Journal of Virology*, 78 (17): 9105–9114.

Lassen, K.G., Ramyar, K.X., Bailey, J.R., Zhou, Y. & Siliciano, R.F. (2006). Nuclear retention of multiple spliced HIV-1 RNA in resting CD4+ T cells . *PLoS Pathogens* 2 (7): e68.

Lehrman, G., Hogue, I.B., Palmer, S., Jennings, C., Spina, C.A., Wiegand, A., Landay, A.L., Coombs, R.W., Richman, D.D., Mellors, J.W., Coffin, J.M., Bosch, R.J. & Margolis, D.M. (2005). Depletion of latent HIV-1 infection in vivo: a proof-of-concept study. *Lancet*, 366: 549-555.

Li, Q., Duan, L., Estes, J.D., Ma, Z.M., Rourke, T., Wang, Y., Reilly, C., Carlis, J., Miller, C.J., Haase, A.T. (2005). Peak SIV replication in resting memory CD4+ T cells depletes gut lamina propria CD4+ T cells. *Nature*, 434 (7037): 1148-1152.

Margolis, D.M. (2010). Treatments for persistent HIV infection: the road ahead. *Retrovirology*, 7 (1): 117.

Pace, M.J., Agosto, L. & O'Doherty, U. (2011). R5 HIV Env and VSV-G Cooperate to Mediate Fusion to Naive CD4+ T Cells. *Journal of Virology*, 85 (1): 644–648.

Pace, M.J., Agosto, L., Graf, E.H. & O'Doherty, U. (2011). HIV reservoirs and latency models. *Virology*, 411: 344-354.

Pasternak, A.O., Jurriaans, S., Bakker, M., Prins, J.M., Berkhout, B. & Lukashov, V.V. (2009). Cellular levels of HIV unspliced RNA from patients on combination antiretroviral therapy with undetectable plasma viremia predict the therapy outcome. *PLoS ONE*, 4 (12): e8490.

Perelson, A.S., Neumann, A.U., Markowitz, M., Leonard, J.M. & Ho, D.D. (1996). HIV-1 dynamics in vivo: virion clearance rate, infected cell life-span, and viral generation time. *Science* 271: 1582-1586.

Perelson, A.S., Essunger, P., Cao, Y., Vesanen, M., Hurley, A., Saksela, K., Markowitz, M. & Ho, D.D. (1997). Decay characteristics of HIV-1-infected compartments during combination therapy. *Nature*, 387: 188-191.

Pierson, T., McArthur, J. & Siliciano, R.F. (2000). Reservoirs for HIV-1: mechanisms for viral persistence in the presence of antiviral immune responses and antiretroviral therapy. *Annual Reviews Immunology*, 18: 665–708.

Plesa, G., Dai, J., Baytop, C., Riley, J.L., June, C.H. & O'Doherty, U. (2007). Addition of deoxynucleosides enhances human immunodeficiency virus type 1 integration and 2LTR formation in resting CD4+ T cells. *Journal of Virology*, 81 (24): 13938–13942.

Reuse, S., Calao, M., Kabeya, K., Guiguen, A., Gatot, J.S., Quivy, V., Vanhulle, C., Lamine, A., Varia, D., Demonte, D., Martinelli, V., Veithen, E., Cherrier, T., Avettand, V., Poutrel, S., Piette, J., de Launoit, Y., Moutschen, M., Burny, A., Rouzioux, C., De

Wit, S., Herbein, G., Rohr, O., Collette, Y., Lambotte, O., Clumeck, N. & Van Lint, C. (2009). Synergistic activation of HIV-1 expression by deacetylase inhibitors and prostratin: implications for treatment of latent infection. *PLoS ONE*, 4: e6093.

Richman, D.D., Margolis, D.M., Delaney, M., Greene, W.C., Hazuda, D. & Pomerantz, R.J. (2009). The challenge of finding a cure for HIV infection. *Science*, 323 (5919): 1304–1307.

Rong, L. & Perelson, A.S. (2009). Modeling HIV persistence, the latent reservoir, and viral blips. *Journal of Theoritical Biology*, 260: 308-331.

Schacker, T.W., Little, S., Connick, E., Gebhard-Mitchell, K., Zhang, Z., Krieger, J., Pryor, J., Havlir, D., Wong, J.K., Richman, D., Corey, L. & Haase, A.T. (2000). Rapid accumulation of human immunodeficiency virus (HIV) in lymphaticc tissue reservoirs during acute and early HIV infection: implications for timing of antiretroviral therapy. *Journal of Infectious Diseases*, 181: 354–357.

Schrager, L.K. & D'Souza, M.P. (1998). Cellular and Anatomical Reservoirs of HIV-1 in Patients Receiving Potent Antiretroviral Combination Therapy. *Journal of the American Medical Association*, 280 (1): 67-71.

Siliciano, J.D. & Siliciano, R.F. (2005). Enhanced culture assay for detection and quantitation of latently infected, resting CD4+ T-cells carrying replication-competent virus in HIV-1-infected individuals. *Methods in Molecular Biolology*, 304: 3–15.

Swingler, S., Brichacek, B., Jacque, J.M., Ulich, C., Zhou, J. & Stevenson, M. (2003). HIV-1 nef intersects the macrophage CD40L signalling pathway to promote resting-cell infection. *Nature*, 424 (6945): 213-219.

Wei X.P., Ghosh S.K., Taylor M.E., Johnson V.A., Emini E.A., Deutsch P, Lifson J.D., Bonhoeffer S., Nowak M.A., Hahn B.H., Saag M.S. & Shaw G.M. (1995). Viral dynamics in human immunodeficiency virus type 1 infection. *Nature*, 373: 117-122.

Williams, S.A. & Greene, W.C. (2007). Regulation of HIV-1 latency by T-cell activation. *Cytokine*, 39: 63-74.

Wong, J.K., Hezareh, M., Günthard, H.F., Havlir, D.V., Ignacio, C.C., Spina, C.A. & Richman, D.D. (1997). Recovery of replication-competent HIV despite prolonged suppression of plasma viremia. *Science*, 278: 1291–1295.

Yang, H.C., Shen, L., Siliciano, R.F. & Pomerantz, J.L. (2009). Isolation of a cellular factor that can reactivate latent HIV-1 without T cell activation. *Proceedings of the National Academy of Sciences USA*, 106: 6321-6326.

Zhang, Z.Q., Wietgrefe, S.W., Li, Q., Shore, M.D., Duan, L., Reilly, C., Lifson, J.D. & Haase, A.T. (2004). Roles of substrate availability and infection of resting and activated CD4+ T cells in transmission and acute simian immunodeficiency virus infection. *Proceedings of the National Academy of Sciences USA*, 101 (15): 5640-5645.

2

Molecular Evolution of Hepatitis Viruses

Flor H. Pujol, Rossana Jaspe and Héctor R. Rangel
Laboratorio de Virología Molecular, CMBC, IVIC, Caracas, Venezuela

1. Introduction

Five hepatitis viruses (HV) are known to date. Infection by enterically-transmitted viruses (HAV and HEV) causes acute hepatitis and is generally benign compared to the disease caused by parenterally-transmitted viruses (HBV, HCV and HDV), for which chronic infection may lead to hepatocellular carcinoma (HCC). Some types of HDV have also been associated to a high frequency of fulminant hepatitis (Table 1). In addition to these viruses, other viruses have been discovered and initially proposed as causative agents of hepatitis, like GBV-C, TTV and SenV. The association with hepatitis was later discarded.

This chapter addresses the molecular evolution of these viruses. An overview on molecular biology and replication of each HV, with emphasis on aspects leading to generation of diversity, is discussed. The diversity of each HV, both at the intrahost and the population level, and the implication of HV diversity on pathogenicity is described. In addition, the possible origin of these viruses is discussed. The chapter also covers how co-infection with HIV may modulate the diversity of HV, in terms of genotypic variability and intrahost evolution.

Hepatitis Virus	**Genome and size**	**Chronicity and HCC**	**Genotypes**	**Salient molecular feature**
A	sRNA 7.5 Kb[1]	No	7: 4 in humans	Codon usage
B	dDNA 3.2 Kb	Yes	8 and simian genotypes	Reverse transcriptase
C	sRNA 9.5 Kb	Yes	7	Quasispecies
D	spRNA 1.7 Kb	Yes	8	Ribozyme, viroid-like genome
E	sRNA 7.2 Kb	No[2]	4	Zoonotic transmission of some genotypes

[1]: s for single stranded, d for partially double stranded, sp for single stranded with intrapairing as a viroid structure. 2: HEV has not been associated to chronicity, except in immunocompromised patients (Kaba et al., 2011).

Table 1. Molecular characteristics of hepatitis viruses

2. Molecular biology and replication of hepatitis viruses

All but one (HBV) of the HV are RNA viruses. This fact implies that they use RNA polymerases - and for HBV a retrotranscriptase - for replication, which lack proofreading capacity, leading to generation of mutations 10^4 more frequently than human DNA polymerase, for example.

2.1 Enterically transmitted viruses

HAV is a non-enveloped virus which belongs to the genus *Hepatovirus*, of the family *Picornaviridae* (Cristina & Costa-Mattioli, 2007). As an RNA virus, replication occurs entirely in the cytoplasm. HAV genome is a single positive stranded ARN of 7.5 Kb, with an Internal Ribosome Entry Side (IRES) at its 5´ non-coding region. It encodes for a polyprotein of aproximately 2.200 aminoacids (Figure 1) (Cristina & Costa-Mattioli, 2007). This polyprotein is cleaved by celular and viral proteases to produce 11 proteins and among them the viral RNA-dependent RNA polymerase. This polymerase produces the antigenomic negative strand RNA which serves as template for generation of genomic positive strand RNAs, which will be inserted into the assembling viral capsids to be liberated by exocytosis throughout the cell (Cuthbert, 2001).

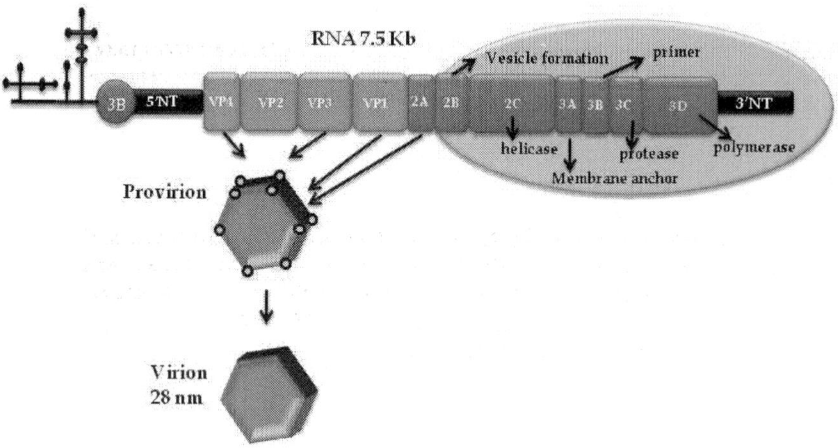

Fig. 1. **HAV virion and genome organization.** The positive-strand RNA genome contains a single open reading frame which encodes a polyprotein proteolytially processed by a cellular protease and the viral protease 3Cpro. Structural proteins (VP) are indicated in green, and nonstructural proteins in blue. The RNA secondary structure of the IRES is shown in the 5´end.

HEV is a non-enveloped virus classified as a *Hepevirus*, in the family *Hepeviridae* (Jameel, 1999). Infection with HEV is responsible for a high percentage of fulminant hepatitis in pregnant women (Dalton et al., 2008). The HEV genome is a positive single-stranded RNA of aproximately 7.2 kb, with a 5´-methylguanine cap and a 3´- polyA stretch. It contains three partially overlapping open reading frames (ORFs) (Figure 2):

- ORF1, coding for non structural proteins including the RNA-dependent RNA polymerase,
- ORF2, coding for the viral capsid protein,

- and ORF3, which might function as a viral accessory protein affecting the host response to infection. (Ahmad et al., 2011).

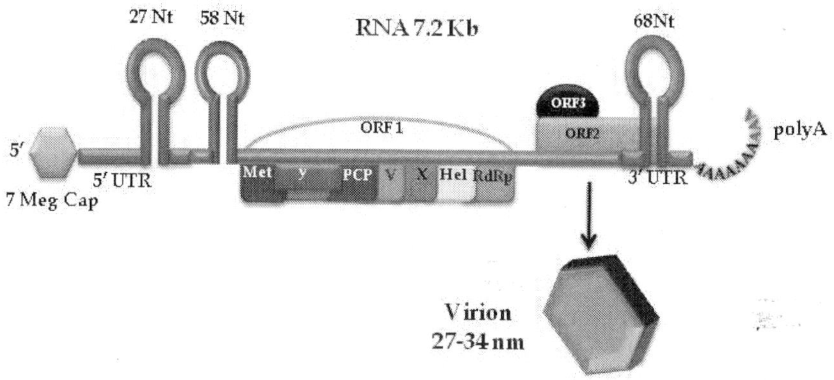

Fig. 2. **HEV virion and genome organization.** The positive strand RNA genome is capped at the 5′ end and polyadenylated at the 3′ end. The three open reading frames (ORFs) are shown. ORF1 encodes the nonstructural polyprotein with various functional units - methyltransferase (MeT), papain-like cysteine protease (PCP), RNA helicase (Hel) and RNA dependent RNA polymerase (RdRp). ORF2 encodes the viral core protein. ORF3 encodes a small regulatory phosphoprotein.

HEV replication is not completely known. After releasing of viral RNA in the cytosol, ORF1 is translated into the polyprotein, generating the replication complex. It is believed that negative RNA intermediates are then produced, for the synthesis of genomic as well as subgenomic positive RNAs, these latter translated into the capsid and ORF3 proteins. Positive genomic RNA is package in the capsids for liberation of HEV through exocytosis (Ahmad et al., 2011).

2.2 Hepatitis B virus

HBV is an enveloped virus belonging to the genus *Hepadnavirus*, in the family *Hepadnaviridae*. This family includes several genera of partially double stranded DNA generated from an intermediate RNA through reverse transcription (Ganem, 1991). HBV genome is around 3,200 bases long (Figure 3), the smallest of all known animal viruses.The viral genome encodes four overlapping ORFs:

- S, coding for the viral surface envelope proteins,
- C, coding for the capsid and e antigen proteins,
- P, coding for the polymerase, functionally divided into the terminal protein domain, which is involved in encapsidation and initiation of minus-strand synthesis; the reverse transcriptase (RT) domain, which catalyzes genome synthesis; and the ribonuclease H domain, which degrades pregenomic RNA.
- X, coding for a protein with multiple functions, including signal transduction, transcriptional activation, DNA repair, and inhibition of protein degradation (Liang, 2009).

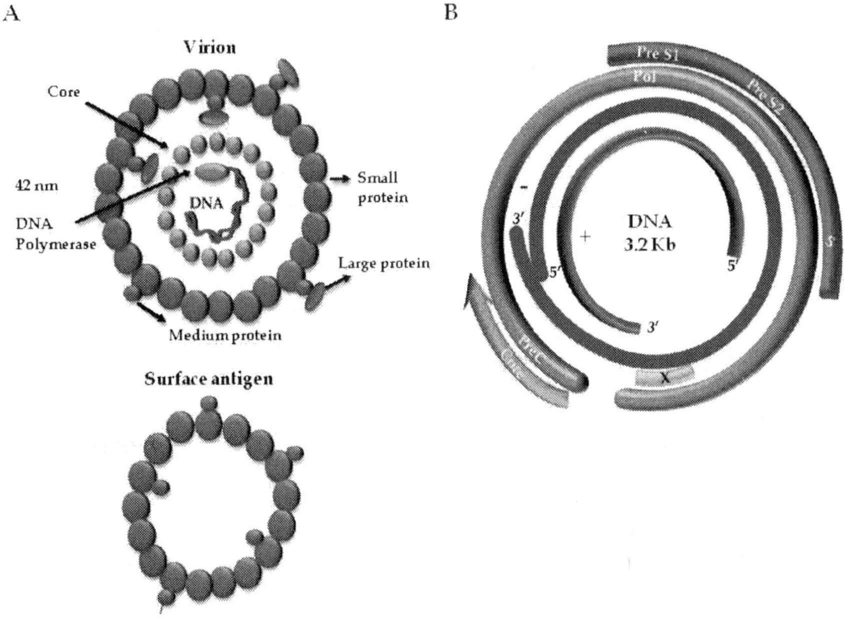

Fig. 3. **HBV virion and genome organization. A.** Enveloped virion of 42 nm. The 3 forms of HBV surface antigen (small, medium ans large) are embedded in the lipid envelop which covers que core which interacts with the partially double stranded DNA, which is covalently linked to the polymerase. Cellular heat shock proteins inside the virion are not shown. In addition to the virion, HBV surface antigen empty particles are secreted both as spherical or cylindrical particles. **B.** The DNA genome and the 4 ORFs are shown, describing the compact nature of this genome.

After entry into the cell, the viral capsids are directed to the nucleus. The single-stranded gap region in the viral genome is repaired and circularized to a covalently closed circular form. This circular DNA is the template for transcription of the pregenomic and several subgenomic messenger RNAs. Pregenomic RNA is retrotranscribed inside the capsids by the HBV polymerase. The nucleocapsids are then directed to the endoplasmic reticulum to interact with the envelope proteins and assemble into mature virions, which are then secreted outside the cell (Liang, 2009).

2.3 Hepatitis D virus

HDV, genus *deltavirus*, is the smallest animal RNA virus (1,700 bases), and is related to plant viroids and satellite viruses (Figure 4). In contrast to plant satellite viruses, HDV is able to perform autounomous replication, but depends on coinfection with HBV, since it uses its viral surface antigen for assembling its virion (Taylor, 2009). Unlike other RNA viruses, HDV lacks an RNA-dependent RNA polymerase, by using the cellular RNA polymerases of the host, which recognize its genome because of its folded, rod-like structure. Three forms of RNA are made in the host during replication: circular genomic and antigenomic RNA, and polyadenylated antigenomic mRNA, which codes for the only protein coded in this genome, the HDAg. Two forms of HDAg are produced by RNA edition. Replication of the circular HDV RNA template occurs via a rolling mechanism similar to that of plant viroids. A viral ribozyme selfcleaves the linear HDV RNA. These monomers are then ligated to form circular RNA, which interacts with HDAg and uses HBV empty surface antigen particles for assembling the HDV virions (Hugues et al., 2011; Taylor, 2009).

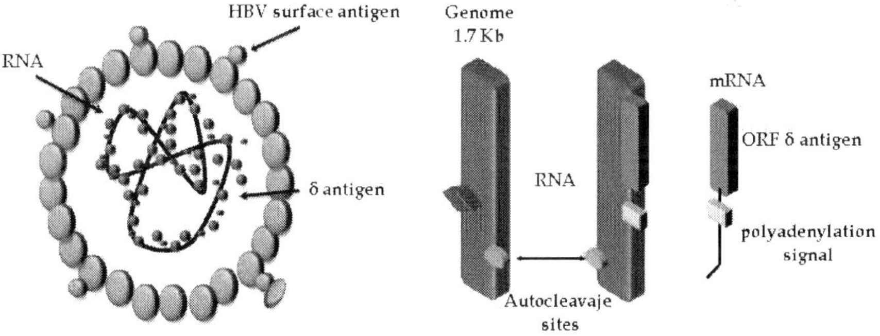

Fig. 4. **HDV virion and genome organization.** HDV RNA is packed through interaction with HDV antigen, who presents in two forms, large (with an N-terminal strech of amino acids intercating with RNA) and small form. HDV RNA and antigens are inside HBV surface antigen empty particles, provided by HBV replication. During replication, 3 HDV RNAs accumulate in the cell: the single negative strand genome and the antigenome, each possessing a ribozyme, with cleavage sites indicated by green circles, and the 800-nucleotide mRNA, that has a 5′-cap and a 3′-poly(A) tail.

2.4 Hepatitis C virus

HCV belongs to the genus *Hepacivirus*, in the family *Flaviviridae*. It is an enveloped virus with a positive RNA of around 9.5 Kb, which codes for a polyprotein of around 3,000 amino acids, including an RNA-dependent RNA polymerase (Figure 5). HCV interacts with a series of receptor to enter the cell via endocytosis, from which the capsid released the viral RNA in a membranous web close to the Endoplasmic reticulum, where replication takes place. The translated polyprotein is co- and post-translationally modified to produce mature viral proteins which can form replication complexes and assemble into new virions. These progeny virions bud into the lumen of the ER and leave the host cell through the secretory pathway (Poenisch & Bartenschlager, 2010; Tang & Grise, 2009).

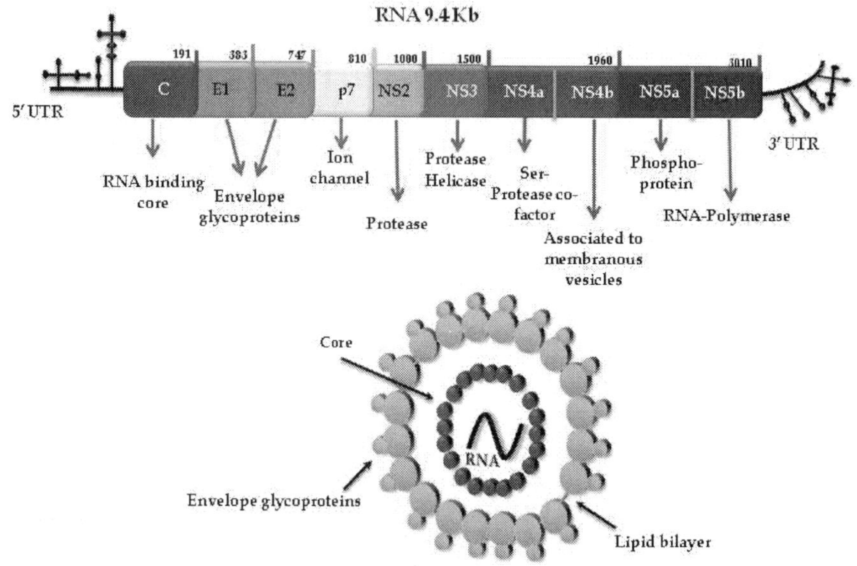

Fig. 5. **HCV virion and genome organization.** The positive-strand RNA genome contains a single open reading frame which encodes a polyprotein proteolytially processed by cellular and viral proteases. Structural proteins (VP) are indicated in green and nonstructural proteins in blue. The RNA secondary structure of the IRES is shown in the 5´end. At the 3´end, the RNA exhbit also a secondary structure involved in replication.

2.5 Other parenterally-transmitted viruses historically associated with hepatitis

In addition to the well established hepatitis viruses,. 3 other parenterally-transmitted viruses have been identified, GBV-C (formerly known as HGV), TTV and SEN-V (Figure 6). However, the real role of these new viruses as causative agents of hepatitis is uncertain. There is no evidence at the present that these viruses cause any pathology in humans (Allain et al., 2002). A tentative genus has been proposed for GBV-C, *Pegivirus*, and 6 genotypes of GBV-C have been described (Smith et al., 2000; Stapleton et al., 2011). Genotype 1 is more prevalent in Africa, genotype 2 in Europe and North America, and genotypes 3, 4 and 5 are found mainly in Asia. GBV-C genotype 3 circulates among Central and South American population groups, a finding that may be related to the Asiatic origin of the American man (Loureiro et al., 2002; Smith et al., 2000). These findings suggest an old origin of GBV-C.

TTV is a single-stranded DNA virus, of around 3.8 Kb, distantly related to circoviruses and classified in the genus Arnellovirus (Biagini, 2009; Hino & Miyata, 2007). TTV represents in fact a swarm of viruses, which chronically infects human and other animals, and was originally thought to be associated with hepatitis. However, there does not seem to be any link between TTV infection and HCC or chronic hepatitis. Up to 23 genotypes of TTV have been described, which are grouped in 5 genogroups. One of these genogroups comprises SEN-V, another virus initially associated with hepatitis. Preliminary evidence suggests that some SEN-

V types may be associated to hepatitis, although this association could be a casual event, and then not meaning that SEN-V is actually a true hepatotropic virus (Hino & Miyata, 2007).

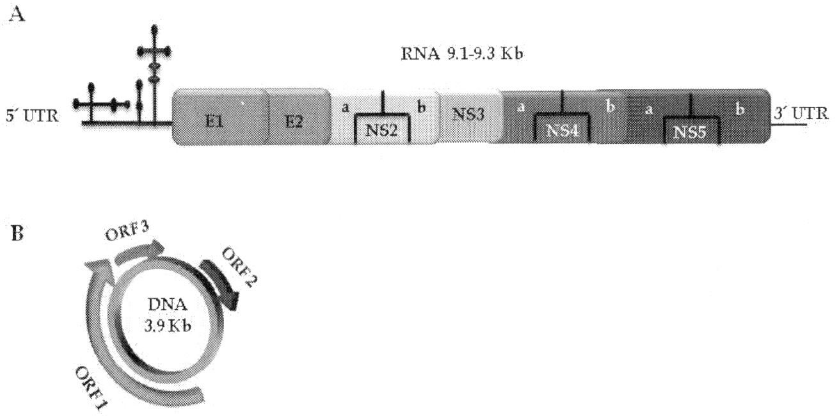

Fig. 6. **Genome organization of GBV-C and Arnellovirus. A: GBV-C genome.** The positive-strand RNA genome contains a single open reading frame which encodes a polyprotein proteolytially processed by cellular and viral proteases. Structural proteins (VP) are indicated in green and nonstructural proteins in blue. A core coding region is absent in the GBV-C genome. The RNA secondary structure of the IRES is shown in the 5´end. **B. Arnellovirus genome.** TTV and Sen-V single-stranded DNA viruses of negative polarity. ORF1 (DNA polymerase), ORF2 (non-structural protein), and ORF3 (core), all present in the plus strand complementary to the virion, are displayed in colours.

3. Genetic diversity of hepatitis viruses

3.1 Enterically transmitted viruses

HAV variants can be classified in 6 genotypes, 3 of them infecting humans and the other 3 other primates from the Old World (Cristina & Costa-Mattioli, 2007; Robertson, 2001) (Table 1). Genotype I, and particularly subgenotype IA, is the most prevalent around the world (Cristina & Costa-Mattioli, 2007). Interestingly, in countries with intermediate to high prevalence of HAV infection, like Latin America and Africa, HAV genotype I is highly predominant, being exclusively found in several countris from South America, where a higher diversity would be expected due to the high frequency of infection (Cristina & Costa-Mattioli, 2007; Sulbarán et al., 2010). A founder effect, like observed for human immunodeficiency virus (HIV) subtype B in the Americas, may account for this situation (Y. Sulbaran et al., 2010; Tebit et al., 2007). In addition, HAV has adopted a naturally highly deoptimized codon usage with respect to that of its cellular host. This characteristic suggests a fine-tuning translation kinetics selection as the underlying mechanism of the codon usage bias in this specific genome region (Aragones et al., 2010). Moreover, significant differences in codon usage are found among the different genotypes (D´Andrea et al., 2011). These differences might be a factor that might bring some adaptative advantage to HAV genotype I and particularly subgenotype IA.

Four genotypes have been reported for HEV (Purcell & Emerson, 2008) (Table 1). Two of these genotypes are endemic among swines and other mammals. An interesting feature of this disease is that two modes of transmission seem to prevail in different geographic regions:

- human to human transmission in highly endemic regions, like Central and Southeat Asia, the Middle East and North Africa, where the most commons genotypes are 1 and 2, the human ones, and
- a zoonotic transmission linked to contact and/or consumption of swines and other susceptible mammals, in non-endemic regions, like Europe, Japan and the Americas, with a more frequent circulation of the animal genotypes 3 and 4 (Purcell & Emerson, 2008). These zoonotic reservoirs might explain the presence of HEV infection in non endemic areas and in isolated populations, like Amerindians, where evidence of exposure to HEV has been documented (Pujol et al., 1994).

3.2 HBV

The absence of proof reading capacity of the HBV reverse transcriptase leads to a high mutation rate. On the other hand, the extreme overlapping of the open reading frames of this small viral genome reduces the viability of many of these mutations (Torresi, 2002). For these opposite characteristics, the substitution rate of HBV is intermediate between RNA and DNA viruses (Kidd-Ljunggren et al., 2002). Another implication of this enhanced potential variability is the generation of a quasispecies-like viral population (Gunther et al., 1999), harboring viral mutations that can be eventually selected under particular selection pressures (Pawlotsky, 2005). The quasispecies complexity is however modulated by the compact genome organization of this virus (Pawlotsky, 2005).

In addition to the diversity which occurs during the natural course of infection, another degree of variability is displayed by HBV strains circulating worldwide. This variability includes the vaccine escape mutants and the genotypic and subtypic variability.

Vaccine escape mutants occur by point mutations in the "a" determinant of the surface antigen, the main immunogenic region, induce conformational changes that prevent the binding of neutralizing antibodies. The most frequent substitutions observed with these characteristics are G145R and D144A (Pawlotsky, 2005; Torresi, 2002). In addition to their transmission between individuals, vaccine escape mutants might be selected under the pressure of neutralizing antibodies or antiretroviral drugs (Lada et al., 2006).

Eight human HBV genotypes (A-H) have been described, based on a minimum divergence of 8% of the complete genome sequences (Table 1) (Figure 7) (Araujo et al., 2011; Norder et al., 2004). Genotypes A and D are predominant in the Old World but are also widely distributed in all the continents. Genotypes B and C are found mainly in South East Asia and the Far East, while genotype E circulates in sub-Saharan West Africa (Norder et al., 2004; Pujol & Devesa, 2005). HBV genotype E might be a recent genotype, exhibitig a low intragenotypic variation not being introduced to the Americas during slave trade (Kramvis et al., 2005; Quintero et al., 2002).

The distribution of genotype G is not fully known. This genotype exhibit several interesting characteristics. A low intragenotypic variability has been found among different isolates from different countries (Lindh, 2005). A high frequency of mutations in the core and precore regions and a frequent association of co-circulation with HBV genotype A have also been reported. This last finding has lead even to the suggestion that the genotype G represents an impaired virus which needs a helper virus for effective replication (Lindh, 2005). However, transmission and infection with exclusively HBVgenotype G has recently been documented (Chudy et al., 2006). On the other hand, a segment of the preS region is identical in genotype E and G strains, suggesting an eventual recombination between these two genotypes. This last assumption might also suggest an African origin for genotype G, although this genotype has not been found yet circulating in Africa (Lindh, 2005). This genotype is might be found more frequently in co-infection with HIV (Dao et al., 2011). Alternatively, HBV genotype G is found frequently infecting men who have sex with men (MSM) (Bottechia et al., 2008; Osiowy et al., 2008; Sanchez et al., 2007). The core variability displayed by HBV genotype G (a 12 amino acid insertion at the N-terminal end) has been shown frecently that migh be affecting the ability of assembly and secretion of the viral particle (Cotelesage et al., 2011), which supports the assumption for the need of a co-infecting strain for an efficient replication of this genotype.

Some of the HBV genotypes are divided into subgenotypes, based on a divergence of more than 4%. Seven subgenotypes are described at the moment for genotype A, 9 for genotype B (Thedja et al., 2011), 12 for genotype C (Mulyanto et al., 2011), 7 for gebnotype D (Meldal et al., 2009) and 4 for genotype F (Devesa et al., 2008). No subgenotypes have been found at present inside genotypes E, G and H. This fact migh be due to the fact that these genotypes might be more recent than the other ones.

HBV genotype F is the most divergent of the HBV genotypes, is autochthonous to and highly predominant in some countries of South America (Devesa & Pujol, 2007). HBV genotype H is closely related to genotype F and seems to be restricted to Central and North America (Arauz-Ruiz et al., 2002). In addition to human HBV genotypes, several simian genotypes have also been identified, one in a monkey from the New World (woolly monkey), while the others have been found infecting simians from the Old World (Figure 7) (Devesa & Pujol, 2007).

In addition, a new genotype I has been proposed for a recombinant of genotypes A, C, and G mainly found in Laos and Vietnam (Tran et al., 2008), genotype J for a recombinant strain between human and ape viruses (Tatematsu et al., 2009). Indeed, several studies have pointed that recombination seems to play an important role in shaping the evolution of HBV (Fares & Holmes, 2002; Simmonds & Midgley, 2005). The exact mechanism of recombination of HBV genomes is not clear, but it seems more likely to occur in the nucleus, by illegitimate replication (Yang & Summers, 1998), or by recombination with integrated HBV DNA (Bowyer & Sim, 2000).

HBV variability seems to play a role in HCC development. Pathogenic differences in causing HCC have been reported among hepatitis B virus (HBV) variants, but also genotypes. HBV genotype C is associated with a more severe disease. (Yang et al., 2008; Yu et al., 2005), and genotype D seems to evolve worse than genotype A (Thakur et al., 2002). HBV genotype F was associated to a higher frequency of HCC development at younger age in Alaskan individuals (Livingston et al., 2007). However, the risk of HCC may differ among subgenotypes (Pujol et al., 2009).

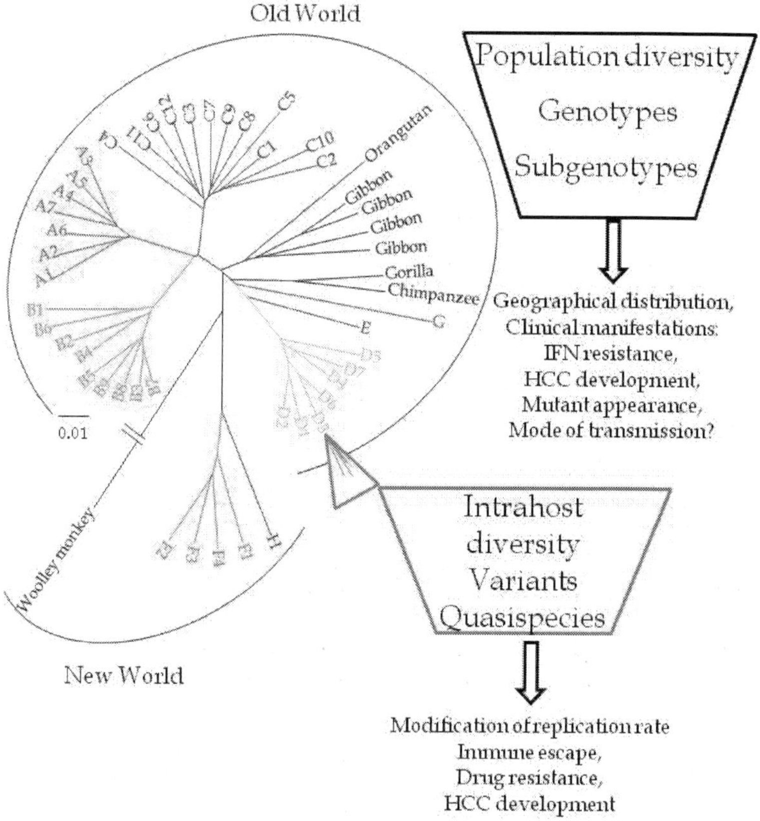

Fig. 7. **HBV molecular evolution.** Human and non-human primates HBV genotypes and subgenotypes are shown in the phylogenetic tree. Intrahost diversity is shown in blue. The implication of HBV variability is also shown.

Several variants are generated during the course of the chronic infection in response to the host or exogenous immune pressures and drug therapy. Mutations in the Precore, Core, X, Pre-S, S and Pol gen, have been reported. Particular interest has been directed toward the generation of translational stop codon mutation at the precore region (mostly G1896A) inside the ε structure, and mutations in the basal core promoter region (especially A1762T, G1764A) and the upstream regulatory sequences (nt1643-1742). The selection of the G1896A seems to be genotype-dependent (Devesa & Pujol, 2007). The basal core promoter overlaps with the X region of the HBV genome, and mutations in the amino acid sequences at positions 130 and 131 in this region (K130M and V131I) has been proposed as prognostic markers for the development of liver cancer (Kuang et al., 2004, Pujol et al., 2009). Some genotypic variability may also occur in terms of interferon sensitivity and development of drug resistance (Kramvis & Kew, 2005; Ramos et al., 2007).

The current treatment for HBV involves the use of Interferon (IFN) and/or antiretroviral drugs, since some of the anti-HIV reverse transcriptase drugs can also inhibits the HBV polymerase. Although no specific mutations have been associated to IFN resistance, some genotypes are more susceptible to this immunomodulator, like genotypes A and B, compared to D and C (Lin & Kao, 2011). Five nucleoside and nucleotide analogues inhibit HBV reverse transcriptase: Adefovir, Entevavir, Lamivudine, Telbivudine and Tenefovir. Drug resistance mutations emerge during treatment with these drugs, consisting of point mutation in one of the 5 domains of the HBV polymerase (Table 2) (Yuen et al., 2009).

The origin of HBV is still an unsolved question (Jazayeri et al., 2009). The reduced size of HBV genome, together with the high degree of overlapping of its open reading frames, has impaired the drawing of an evolutionary picture of this virus. With the advent of sequences from several HBV strains circulating in non human primates (Figure 7), an alternative hypothesis has been proposed: human HBV genotypes might have emerged through several zoonotic introductions from simian strains, both at the Old and New World (Devesa & Pujol, 2007).

3.3 HDV

Eight genotypes of HDV have been identified (Table 1) (Deny, 2006). HDV genotype 1 is present worldwide. Genotype 2 is found in Japan, Taiwan, Russia. Genotype 3 is the most divergent genotype and is found in the Amazon Basin, and has been shown to infect individuals from Peru, Venezuela and Colombia, where severe cases have been documented. This genotype is actually the most frequently associated to fulminant hepatitis. Genotype 4 circulates in Taiwan and Japan. The remaining HDV genotypes (5–8) are found Africa (Deny, 2006).

As other RNA viruses, HDV circulates as a quasispecies distribution of variants, in which defective mutants have being described (Wu et al., 2005). In addition mutants appearing under the immune pressure, as detected by the presence of amino acids under positive selection, target of cytotoxic T lymphocytes, have also been described (Wang et al., 2007).

3.4 HCV

HCV has been classified in 7 genotypes, according to a genetic divergence of more than 30-35% in the complete genome and in several subtypes inside each genotype, according to divergences of more than 20% (Figure 8) (Le Guillou-Guillemette et al., 2007; Chayama & Hayes, 2011). Infections with HCV genotype 1 are associated with the lowest therapeutic success (Zeuzem et al., 2000). HCV genotypes 1, 2, and 3 have a worldwide distribution. HCV subtypes 1a and 1b are the most common genotypes in the US and are also are predominant in Europe, while in Japan, subtype 1b is predominant. Although HCV subtypes 2a and 2b are relatively common in America, Europe, and Japan, subtype 2c is found commonly in northern Italy. HCV genotype 3a is frequent in intravenous drug abusers in Europe and the United States. HCV genotype 4 is prevalent in Africa and the Middle East, and genotypes 5 and 6 seem to be confined to South Africa and Asia, respectively (Simonds, 2001; Zein, 2000). HCV genotype 7 was more recently identified in Canada, in an emigrant from the Democratic Republic of Congo.

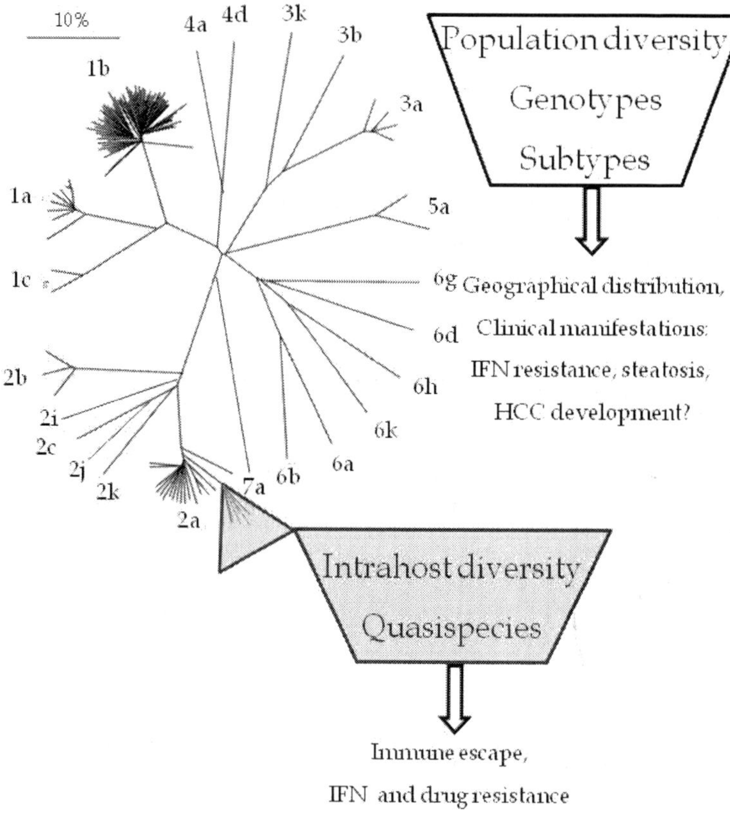

Fig. 8. **HCV molecular evolution**. HCV genotypes and subtypes for which at least one complete genome has been sequenced, are shown in the phylogenetic tree. Intrahost diversity inside each subtype is shown in blue. The implication of HCV variability is also shown.

In contrast with HBV, recombination between HCV genotypes seems to be a rare event. *In vitro* studies suggest a low frequency of recombination for this virus (Reiter et al., 2011), in agreement with the low number of recombinant strains identified so far (Morel et al., 2011). However, intergenotypic incompatibility might be a factor involved in the low frequency of recombinants observed, and intragenotypic recombination might be more frequent than expected (Mes & Doomum, 2011).

Changes in hepatitis C virus (HCV) genotype distribution with time have been reported in several countries. In Venezuela, for example, a significant reduction of the circulation of HCV genotype 1b was observed in the last decade, with the increase of circulation of genotype 2j (Pujol & Loureiro, 2007; M.Z. Sulbaran et al., 2010). Several subtypes of HCV genotype 2 and 4 were introduced in some countries of the Americas during slave trade in

Martinique (Martial et al., 2004) and of HCV genotype 2 in Venezuela (Sulbarán M.Z. et al., 2010). It is difficult to estimate for how long HCV has been present in human populations. HCV may have been endemic in Asia and Africa for a considerably longer time than in Western countries. HCV subtypes might have diverged around 200-250 years ago and genotypes around 500-2000 years ago (Bostan & Mahmood, 2010; Cantaloube et al., 2003; Smith et al., 1997).

HCV genotype 1b has been frequently associated with a more severe liver disease. Nevertheless, this association might be due to the fact that individuals infected with this genotype have a longer mean duration of infection (Zein, 2000). A recent meta-analysis showed however HCV subtype 1b associated to a higher risk factor for HCC development (Raimondi et al., 2009). Hepatic steatosis is a common consequence of HCV infection, has been recently associated with the development of HCC, and is more frequently found among HCV genotype 3 infected patients (Monto et al., 2002). More studies are needed to confirm the correlation between HCV genotype 3, the presence of steatosis and progression to HCC (Zhu & Chung, 2003).

Within an infected individual, HCV circulates as a collection of closely related variants named quasispecies (Fishman & Branch, 2009). HCV like others RNA virus has a high level of genetic variability, especially in the E1 and E2 genes (envelope glycoproteins). Humoral pressure on the hypervariably region of E2 has been associated with quasispecies diversification through immune escape mechanisms (Weiner et al., 1992). The quasispecies nature of HCV populations in an infected individual might contribute to the viral persistence in the host, IFN and drug resistance (Fishman & Branch, 2009) (Figure 8).

4. HIV co-infection

HIV co-infection with HCV and/or HBV is frequent, since these viruses share many modes of transmission, and exacerbates the natural history of these viral infections. A decreased immune clearance and more rapid progression of liver disease has been documented, leading to an increased incidence of cirrhosis, risk of drug-related hepatoxicity, HCC development, and death. On the other hand, liver disease has emerged as a major cause of morbidity and mortality in HIV-infected patients. Viral hepatitis co-infection increases the risk of drug-related hepatoxicity of highly active antiretroviral therapy (HAART), impacting the selection of specific agents (Sulkowski, 2008).

HIV-1 co-infection increases HCV viral load in dually infected patients. This effect seems to be both related to the acquired immunodeficiency and to a direct interaction between the viruses (Rotman & Liang, 2009). In addition, HIV-1 co-infection seems to impact the quasispecies complexity exhibited by HCV. Both an increase and a decrease of quasispecies heterogeneity has been described, when compared to HCV mono-infected patients (Sherman et al., 1996; Toyoda et al., 1997). In spite of the conflicting results, many reports suggest a relatively low level of HCV quasispecies diversity before HAART, and an increase in diversity after prolonged treatment, principally due to a significant increase in both synonymous and non-synonymous substitution rates in the hypervariable region 1 of HCV E2 (Bernini, et al, 2011). This exponential growth of the quasispecies populations in immunological responders coincides with a peak in CD4 cell counts, positive selection in several proteins of HCV, and a frequent increase in HCV viral load (Bernini, et al, 2011).

Patients co-infected with HBV and HIV-1 have a higher likelihood of chronicity after acute HBV infection compared with HIV-negative patients (Lacombe et al., 2010). The natural history of HBV-related disease is modified by HIV infection in several ways. Co-infected patients have higher HBV DNA levels, lower aminotransferase levels, decreased spontaneous loss of hepatitis B early antigen (HBeAg), accelerated progression to cirrhosis, and increased risk of liver-related morbidity and mortality compared with HBV monoinfection (Lacombe et al., 2010). As previously mentioned, a number of clinically significant HBV genome mutations have been reported in HBV mono-infection, and differences in disease evolution and treatment response have been associated to a particular genotype and to some of these mutations. Many of these mutations appears during the long term evolution of infection and during exposure to nucleos(t)ides analogs, used to treat HIV (Lacombe et al., 2010). In HIV-1/HBV co-infected individuals, a novel –1G mutation in the HBV core and precore gene was found to be more frequent compared to mono-infected patients. This mutation results in premature termination of the deduced HBV precore and core genes and was associated with high HBV viral load. PreS2 deletions were observed more frequently in co-infection (Audsley et al., 2010).

An interesting observation is that the natural course of HIV-1 might be modulated by the presence of GBV-C. GBV-C co-infection seems to exert a beneficial effect on HIV disease progression (Maidana et al., 2005), although this evidence has not been consistently corroborated (Baggio-Zappia et al., 2009). The mechanism by which GBV-C interferes with AIDS progression is not yet fully undestood. *In vitro* studies have shown that the inhibitory effect of GBV-C on HIV-1 replication might be related to solubles factors induced by GBV-C (Jung, 2005), and specifically with chemokines, Rantes, MIP-1 and SDF-1, which may compete with HIV-1 envelop glycoprotein 120 for the correceptor CCR5 and CXCR4 in CD4 cells. This effect may reduce a successful HIV-1 interaction with infected cell (Xiang, 2004). Several lines of evidence suggest a possible inhibition of HIV-1 by GBV-C through increase in soluble ligands for HIV-1 correceptor and activation of innate immunity (Lalle, 2008; Mohr & Stapleton, 2009). In addition, the beneficial effect of GBV-C on survival of HIV-1 infected patients might be genotype specific (Schwarze-Zander et al., 2006). Two GBV-C proteins, E2 and NS5A, have been shown to modulate CD4+ T-lymphocyte chemokine receptor expression and chemokine release in vitro, then inhbiting HIV replication. The inhibitory effect of GBV-C NS5A on HIV-1 replication was exerced *in vitro* by the non structural proteins from all the genotypes tested (Chang et al., 2007), failing then to describe a genotype specific inhibition of this protein in HIV-1 replication. More studies are needed to clarify the exact role of GBV-C co-infection on HIV-1 replication.

5. Conclusions

There are still 5 viral entities named hepatitis viruses, which share only their tropism for the hepatocyte. Parenterally transmitted viruses are normally the ones associated to chronicity and to more severe sequela, like cirrhosis and HCC. Due to the error prone nature of their polymerases, these viruses display a substantial degree of genetic diversity. Within these 5 viral entities, viral variants (genotypes, subgenotypes, diversity of quasispecies, mutants) might exhibit particular characteristics in term of pathogenesis and mode of transmission. Thus, instead of 5 viruses, we are in fact dealing with a multiplicity of viral variants with different consequences and evolution inside the infected host. Vaccines are not available for all these entities, only for HAV and HBV, and partially for HDV. The significant degree of variability exhibited by these viruses is an unresolved limitation for the development of effective vaccines against them. Some of these variants might have originated separately in the New and the Old World, as for HBV and HDV, for example. Some of these viruses may have a long time of co-evolution with human host, as for GBV-C, while others might have been introduced more recently, like HCV.

6. Acknowledgment

We wish to thank Denisse Guevara for her help with the design of figures.

7. References

Ahmad, I., Holla, R.P. & Jameel, S. (2011). Molecular virology of hepatitis E virus. *Virus Research*, in press. ISSN: 0168-1702

Allain, J.P., Thomas, I. & Sauleda, S. (2002). Nucleic acid testing for emerging viral infections. *Transfusion Medicine*, 12(4): 275-283. ISSN: 1365-3148

Aragonès, L., Guix, S., Ribes, E., Bosch, A. & Pintó, R.M. (2010). Fine-tuning translation kinetics selection as the driving force of codon usage bias in the hepatitis A virus capsid. *PLoS Pathogen*, 6: e1000797. ISSN: 1553-7374

Araujo, N.M., Waizbort, R. & Kay, A. (2011). Hepatitis B virus infection from an evolutionary point of view: How viral, host, and environmental factors shape genotypes and subgenotypes. *Infection Genetics and Evolution*, 11(6): 1199-1207. ISSN: 1567-1348

Arauz-Ruiz, P., Norder, H., Robertson, B.H. & Magnius, L.O. (2002). Genotype H: a new Amerindian genotype of hepatitis B virus revealed in Central America. *Journal of General Virology*, 83(Pt 8): 2059-2073. ISSN: 1465-2099

Audsley, J., Littlejohn, M., Yuen, L., Sasadeusz, J., Ayres, A., Desmond, C., Spelman, T., Lau, G., Matthews, G.V., Avihingsanon, A., Seaberg, E., Philp, F., Saulynas, M., Ruxrungtham, K., Dore, G.J., Locarnini, S.A., Thio, C.L., Lewin, S.R. & Revill, P.A. (2010). HBV mutations in untreated HIV-HBV co-infection using genomic length sequencing. *Virology*, 405(2): 539-547. ISSN: 0042-6822

Baggio-Zappia, G.L. & Hernandes Granato, C.F. (2009). HIV-GB virus C co-infection: an overview. *Clinical Chemistry and Laboratory Medicine*, 47: 12-19. ISSN: 1437- 4331

Biagini, P. (2009). Classification of TTV and related viruses (anelloviruses). *Current Topics in Microbiology and Immunology*, 331: 21-33. ISSN: 0070-217X

Bernini, F., Ebranati, E., De Maddalena, C., Shkjezi, R., Milazzo, L., Lo Presti, A., Ciccozzi, M., Galli, M. & Zehender, G. (2011). Within-host dynamics of the hepatitis C virus quasispecies population in HIV-1/HCV coinfected patients. *PLoS One*. 6(1): e16551. ISSN: 1932-6203

Bostan, N. & Mahmood, T. (2010). An overview about hepatitis C: a devastating virus. *Critical Reviews in Microbiology*, 36(2): 91-133. ISSN: 1549-7828

Bottecchia, M., Souto, F.J., O, K.M., Amendola, M., Brandão, C.E., Niel, C. & Gomes, S.A. (2008). Hepatitis B virus genotypes and resistance mutations in patients under long term lamivudine therapy: characterization of genotype G in Brazil. *BMC Microbiology*, 8: 11. ISSN 1471-2180

Bowyer, S.M. & Sim, J.G. (2000). Relationships within and between genotypes of hepatitis B virus at points across the genome: footprints of recombination in certain isolates. *Journal of General Virology*, 81(Pt 2): 379-392. ISSN: 1465-2099

Cantaloube, J.F., Biagini, P., Attoui, H., Gallian, P., de Micco, P. & de Lamballerie, X. (2003). Evolution of hepatitis C virus in blood donors and their respective recipients *Journal of General Virology*, 84(Pt 2): 441-446. ISSN: 1465-2099

Chang, Q., McLinden, J.H., Stapleton, J.T., Sathar, M.A. & Xiang, J. (2007). Expression of GB virus C NS5A protein from genotypes 1, 2, 3 and 5 and a 30 aa NS5A fragment inhibit human immunodeficiency virus type 1 replication in a CD4+ T-lymphocyte cell line. *Journal of General Virology*, 88(Pt 12):3341-3346. ISSN: 1465-2099

Chayama, K. & Hayes, C.N. (2011). Hepatitis C virus: How genetic variability affects pathobiology of disease. *Journal of Gastroenterology and Hepatology*, 26(Suppl 1): 83-95. ISSN: 1440-1746

Chudy, M., Schmidt, M., Czudai, V., Scheiblauer, H., Nick, S., Mosebach, M., Hourfar, M.K., Seifried, E., Roth, W.K., Grünelt, E. & Nübling, C.M. (2006). Hepatitis B virus genotype G monoinfection and its transmission by blood components. *Hepatology*, 44(1): 99-107. ISSN: 1527-3350

Cotelesage, J.J., Osiowy, C., Lawrence, C., DeVarennes, S.L., Teow, S., Beniac, D.R. & Booth, T.F. (2011). Hepatitis B Virus Genotype G forms core-like particles with unique structural properties. *Journal of Viral Hepatitis*, 18(6): 443-448. ISSN: 1365-2893

Cristina, J. & Costa-Mattioli, M. (2007). Genetic variability and molecular evolution of hepatitis A virus. *Virus Research*, 127(2): 151-157. ISSN: 0168-1702

Cuthbert, J.A. (2001). Hepatitis A: old and new. *Clinical Microbiology Reviews*, 14(1): 38-58. ISSN: 1098-6618

Dao, D.Y., Balko, J., Attar, N., Neak, E., Yuan, H.J., Lee, W.M. & Jain, M.K. (2011). Hepatitis B virus genotype G: prevalence and impact in patients co-infected with human immunodeficiency virus. *Journal of Medical Virology*, 83(9): 1551-1558. ISSN: 1096-9071

Dalton, H.R., Bendall, R., Ijaz, S. & Banks, M. (2008). Hepatitis E: an emerging infection in developed countries. *Lancet Infectious Diseases*, 8(11): 698-709. ISSN: 1473-3099

D' Andrea, L., Pintó, R.M., Bosch, A., Musto, H. & Cristina, J. (2011). A detailed comparative analysis on the overall codon usage patterns in hepatitis A virus. *Virus Research*, 157(1): 19-24. ISSN: 0168-1702

Dény, P. (2006). Hepatitis delta virus genetic variability: from genotypes I, II, III to eight major clades? *Current Topics in Microbiology and Immunology*, 307: 151-171. ISSN: 0070-217X

Devesa, M. & Pujol, F.H. (2007). Hepatitis B virus genetic diversity in Latin America. *Virus Research*, 127(2): 177-184. ISSN: 0168-1702

Devesa, M., Loureiro, C.L., Rivas, Y., Monsalve, F., Cardona, N., Duarte, M.C., Poblete, F., Gutierrez, M.F., Botto, C. & Pujol, F.H. (2008). Subgenotype diversity of hepatitis B virus American genotype F in Amerindians from Venezuela and the general population of Colombia *Journal of Medical Virology*, 80(1): 20-26. ISSN: 1096- 9071

Fares, M.A. & Holmes, E.C. (2002). A revised evolutionary history of hepatitis B virus (HBV). *Journal of Molecular Evolution*, 54(6): 807-814. ISSN: 1432-1432

Fishman, S.L. & Branch, A.D. (2009). The quasispecies nature and biological implications of the hepatitis C virus. *Infection Genetics and Evolution*, 9(6): 1158-1167. ISSN: 1567-1348

Ganem, D. (1991). Assembly of hepadnaviral virions and subviral particles. *Current Topics in Microbiology and Immunology*, 168: 61-83. ISSN: 0070-217X

Gunther, S., Fischer, L., Pult, I., Sterneck, M. & Will, H. (1999). Naturally occurring variants of hepatitis B virus. *Advances in Virus Research*, 52: 25-137. ISSN: 0065-3527

Hino, S. & Miyata, H. (2007). Torque teno virus (TTV): current status. *Reviews in Medical Virology*, 17(1): 45-57. ISSN: 1099-1654

Hughes, S.A., Wedemeyer, H. & Harrison, P.M. (2011). Hepatitis delta virus. *Lancet*, 378(9785): 73-85. ISSN: 0140-6736

Jameel, S. (1999). Molecular biology and pathogenesis of hepatitis E virus. *Expert Reviews in Molecular Medicine*, 1999: 1-16. ISSN:1462-3994

Jazayeri, S.M., Alavian, S.M. & Carman, W.F. (2010). Hepatitis B virus: origin and evolution. *Journal of Viral Hepatitis*, 17(4): 229-235. ISSN: 1365-2893

Jung, S., Knauer, O., Donhauser, N., Eichenmüller, M., Helm, M., Fleckenstein, B. & Reil, H. (2005). Inhibition of HIV strains by GB virus C in cell culture can be mediated by CD4 and CD8 T-lymphocyte derived soluble factors. *AIDS*, 19(12): 1267-1272. ISSN: 1473-5571

Kaba, M., Richet, H., Ravaux, I., Moreau, J., Poizot-Martin, I., Motte, A., Nicolino-Brunet, C., Dignat-George, F., Ménard, A., Dhiver, C., Brouqui, P. & Colson, P. (2011). Hepatitis E virus infection in patients infected with the human immunodeficiency virus. *Journal of Medical Virology*, 83(10): 1704-1716. ISSN: 1096- 9071

Kidd-Ljunggren, K., Miyakawa, Y. & Kidd, A.H. (2002). Genetic variability in hepatitis B viruses. *Journal of General Virology*, 83(Pt 6): 1267-1280. ISSN: 1465-2099

Kramvis, A. & Kew, M.C. (2005). Relationship of genotypes of hepatitis B virus to mutations, disease progression and response to antiviral therapy *Journal of Viral Hepatitis*, 12(5): 456-464. ISSN: 1365-2893

Kramvis, A., Restorp, K., Norder, H., Botha, J.F., Magnius, L.O. & Kew, M.C. (2005). Full genome analysis of hepatitis B virus genotype E strains from South-Western Africa and Madagascar reveals low genetic variability. *Journal of Medical Virology*, 77(1): 47-52. ISSN: 1096- 9071

Kuang, S.Y., Jackson, P.E., Wang, J.B., Lu, P.X., Muñoz, A., Qian, G.S., Kensler, T.W. & Groopman, J.D. (2004). Specific mutations of hepatitis B virus in plasma predict liver cancer development. *Proceedings of the National Academy of Sciences* USA, 101(10): 3575-3580. ISSN: 1091- 6490

Lacombe, K., Bottero, J., Lemoine, M., Boyd, A. & Girard, P.M. (2010). HIV/hepatitis B virus co-infection: current challenges and new strategies. *Journal of Antimicrobial Chemotherapy*, 65(1): 10-17. ISSN: 0305-7453

Lada, O., Benhamou, Y., Poynard, T. & Thibault, V. (2006). Coexistence of hepatitis B surface antigen (HBs Ag) and anti-HBs antibodies in chronic hepatitis B virus carriers: influence of "a" determinant variants. *Journal of Virology*, 80(6): 2968-2975. ISSN: 1098-5514

Lalle, E., Sacchi, A., Abbate, I., Vitale, A., Martini, F., D'Offizi, G., Antonucci, G., Castilletti, C., Poccia, F. & Capobianchi, M.R. (2008). Activation of interferon response genes and of plasmacytoid dendritic cells in HIV-1 positive subjects with GB virus C co-infection. *International Journal of Immunopathology & Pharmacology*, 21(1): 161–171. ISSN: 0394-6320

Le Guillou-Guillemette, H., Vallet, S., Gaudy-Graffin, C., Payan, C., Pivert, A., Goudeau, A. & Lunel-Fabiani, F. (2007). Genetic diversity of the hepatitis C virus: impact and issues in the antiviral therapy. *World Journal of Gastroenterology*, 13(17): 2416-2426. ISSN: 1007-9327

Liang, T.J. (2009). Hepatitis B: the virus and disease. *Hepatology*, 49(5 Suppl): S13-S21. ISSN: 1527-3350

Lin, C.L. & Kao, J.H. (2011). Recent advances in the treatment of chronic hepatitis B. *Expert Opinion on Pharmacotherapy*, 12(13): 2025-2040. ISSN: 1744-7666

Lindh, M. (2005). HBV genotype G-an odd genotype of unknown origin. *Journal of Clinical Virology*, 34(4): 315-316. ISSN : 1386-6532

Livingston, S.E., Simonetti, J.P., McMahon, B.J., Bulkow, L.R., Hurlburt, K.J., Homan, C.E., Snowball, M.M., Cagle, H.H., Williams, J.L. & Chulanov, V.P. (2007). Hepatitis B virus genotypes in Alaska Native people with hepatocellular carcinoma: preponderance of genotype F. *Journal of Infectious* Diseases, 195(1): 5-11. ISSN: 1537-6613

Loureiro, C.L., Alonso, R., Pacheco, B.A., Uzcátegui, M.G., Villegas, L., León, G., De Saéz, A., Liprandi, F., López, J.L. & Pujol, F.H. (2002). High prevalence of GB virus C/hepatitis G virus genotype 3 among autochthonous Venezuelan populations. *Journal of Medical Virology*, 68(3): 357-362. ISSN: 1096- 9071

Maidana, M.T., Sabino, E.C. & Kallas, E.G. (2005). GBV-C/HGV and HIV-1 coinfection. *Brazilian Journal of Infectious Diseases*, 9(2): 122-125. ISSN 1413-8670

Martial, J., Morice, Y., Abel, S., Cabié, A., Rat, C., Lombard, F., Edouard, A., Pierre-Louis, S., Garsaud, P., Béra, O., Chout, R., Gordien, E., Deny, P. & Césaire, R. (2004). Hepatitis C virus (HCV) genotypes in the Caribbean island of Martinique: evidence for a large radiation of HCV-2 and for a recent introduction from Europe of HCV-4. *Journal of Clinical Microbiology*, 42(2): 784-791. ISSN: 1098-660X

Meldal, B.H., Moula, N.M., Barnes, I.H., Boukef, K. & Allain, J.P. (2009). A novel hepatitis B virus subgenotype, D7, in Tunisian blood donors. *Journal of General Virology*, 90(Pt 7): 1622-1628. . ISSN: 1465-2099

Mes, T.H. & van Doornum, G.J. (2011). Recombination in hepatitis C virus genotype 1 evaluated by phylogenetic and population-genetic methods *Journal of General Virology*, 92(Pt 2): 279-286. . ISSN: 1465-2099

Mohr, E.L. & Stapleton, J.T. (2009). GB virus type C interaction with HOV: the role of envelope glycoproteins. *Journal of Viral Hepatitis*, 16(11): 757-768. ISSN: 1365-2893

Monto, A. (2002). Hepatitis C and steatosis. *Seminars in gastrointestinal disease*, 13(1): 40-46. ISSN:1049-5118

Morel, V., Fournier, C., François, C., Brochot, E., Helle, F., Duverlie, G. & Castelain, S. (2011). Genetic recombination of the hepatitis C virus: clinical implications. *Journal of Viral Hepatitis*, 18(2): 77-83. ISSN: 1365-2893

Mulyanto, Depamede, S.N., Wahyono, A., Jirintai, Nagashima, S., Takahashi, M, & Okamoto, H. (2011). Analysis of the full-length genomes of novel hepatitis B virus subgenotypes C11 and C12 in Papua, Indonesia. *Journal of Medical Virology*, 83(1): 54-64. . ISSN: 1096- 9071

Norder, H., Courouce, A.M., Coursaget, P., Echevarria, J.M., Lee, S.D., Mushahwar, I.K., Robertson, B.H., Locarnini, S. & Magnius, L.O. (2004). Genetic diversity of hepatitis B virus strains derived worldwide: genotypes, subgenotypes, and HBsAg subtypes. *Intervirology*, 47(6): 289-309. ISSN: 1423-0100

Osiowy, C., Gordon, D., Borlang, J., Giles, E. & Villeneuve, J.P. (2008). Hepatitis B virus genotype G epidemiology and co-infection with genotype A in Canada *Journal of General Virology*, 89(Pt 12): 3009-3015. ISSN: 1465-2099

Pawlotsky, J.M. (2005). The concept of hepatitis B virus mutant escape. *Journal of Clinical Virology*, 34(Suppl 1): S125-S129. ISSN : 1386-6532

Poenisch, M. & Bartenschlager, R. (2010). New insights into structure and replication of the hepatitis C virus and clinical implications. *Seminars in Liver Disease*, 30(4): 333-347. ISSN: 0272-8087

Pujol, F.H., Favorov, M.O., Marcano, T., Esté, J.A., Magris, M., Liprandi, F., Khudyakov, Y.E., Khudyakova, N.S. & Fields, H.A. (1994). Prevalence of antibodies against hepatitis E virus among urban and rural population in Venezuela. *Journal of Medical Virology*, 42(3): 234-236. ISSN: 1096- 907

Pujol, F.H. & Devesa, M. (2005). Genotypic variability of hepatitis viruses associated with chronic infection and the development of hepatocellular carcinoma. *Journal of Clinical Gastroenterology*, 39(7): 611-618. ISSN: 0192-0790

Pujol, F.H. & Loureiro, C.L. (2007). Replacement of hepatitis C virus genotype 1b by genotype 2 over a 10-year period in Venezuela. *Journal of Clinical Gastroenterology*, 41(5): 518-520. ISSN: 0192-0790

Pujol, F.H., Navas, M.C., Hainaut, P. & Chemin, I. (2009) Worldwide genetic diversity of HBV genotypes and risk of hepatocellular carcinoma. *Cancer Letters*, 286(1): 80-88. ISSN: 0304-3835

Purcell, R.H. & Emerson, S.U. (2008). Hepatitis E: an emerging awareness of an old disease. *Journal of Hepatology*, 48(3): 494-503. ISSN: 0168-8278

Quintero, A., Martinez, D., Alarcon De Noya, B., Costagliola, A., Urbina, L., Gonzalez, N., Liprandi, F., Castro De Guerra, D. & Pujol, F.H. (2002). Molecular epidemiology of hepatitis B virus in Afro-Venezuelan populations. *Archives of Virology*, 147(9): 1829-1836. ISSN:0304-8608

Raimondi, S., Bruno, S., Mondelli, M.U. & Maisonneuve, P. (2009). Hepatitis C virus genotype 1b as a risk factor for hepatocellular carcinoma development: a meta-analysis. *Journal of Hepatology*, 50(6): 1142-1154. ISSN: 0168-8278

Ramos, B., Núñez, M., Martín-Carbonero, L., Sheldon, J., Rios, P., Labarga, P., Romero, M., Barreiro, P., García-Samaniego, J. & Soriano, V. (2007). Hepatitis B virus genotypes and lamivudine resistance mutations in HIV/hepatitis B virus-coinfected patients. *Journal of Acquired Immune Deficiency Syndromes*, 44(5): 557-561. ISSN: 1525-4135

Reiter, J., Pérez-Vilaró, G., Scheller, N., Mina, L.B., Díez, J. & Meyerhans, A. (2011). Hepatitis C virus RNA recombination in cell culture. *Journal of Hepatology*, in press. ISSN: 0168-8278

Robertson, B.H. (2001). Viral hepatitis and primates: historical and molecular analysis of human and nonhuman primate hepatitis A, B, and the GB-related viruses. *Journal of Viral Hepatitis*, 8(4): 233-242. ISSN: 1365-2893

Rotman, Y. & Liang, T.J. (2009). Coinfection with hepatitis C virus and human immunodeficiency virus: virological, immunological, and clinical outcomes. *Journal of Virology,* 83(15): 7366-7374. ISSN: 1098-5514

Sánchez, L.V., Tanaka, Y., Maldonado, M., Mizokami, M. & Panduro, A. (2007). Difference of hepatitis B virus genotype distribution in two groups of mexican patients with different risk factors. High prevalence of genotype H and G. *Intervirology,* 50(1): 9-15. ISSN: 1423-0100

Schwarze-Zander, C., Blackard, J.T., Zheng, H., Addo, M.M., Lin, W., Robbins, G.K., Sherman, K.E., Zdunek, D., Hess, G., Chung, R.T. & AIDS Clinical Trial Group A5071 Study Team. (2006). GB virus C (GBV-C) infection in hepatitis C virus (HCV)/HIV-coinfected patients receiving HCV treatment: importance of the GBV-C genotype. *Journal of Infectious Diseases,* 194(4): 410-419. ISSN: 0022-1899

Sherman, K.E., Andreatta, C., O'Brien, J., Gutierrez, A. & Harris, R. (1996). Hepatitis C in human immunodeficiency virus-coinfected patients: increased variability in the hypervariable envelope coding domain. *Hepatology,* 23(4): 688–694. ISSN: 1527-3350

Simmonds, P. (2001). Reconstructing the origins of human hepatitis viruses. *Philosophical Transactions of the Royal Society of London. Series B, Biological Sciences,* 356(1411): 1013-1026. ISSN:0022-5193

Simmonds, P. & Midgley, S. (2005). Recombination in the genesis and evolution of hepatitis B virus genotypes. *Journal of Virology,* 79(24): 15467-15476. ISSN: 1098-5514

Smith, D.B., Pathirana, S., Davidson, F., Lawlor, E., Power, J., Yap, P.L. & Simmonds, P. (1997). The origin of hepatitis C virus genotypes. *Journal of General Virology,* 78(Pt 2): 321-328. ISSN: 1465-2099

Smith, D.B., Basaras, M., Frost, S., Haydon, D., Cuceanu, N., Prescott, L., Kamenka, C., Millband, D., Sathar, M.A. & Simmonds, P. (2000). Phylogenetic analysis of GBV-C/hepatitis G virus. *Journal of General Virology,* 81(Pt 3): 769-780. ISSN: 1465-2099

Stapleton, J.T., Foung, S., Muerhoff, A.S., Bukh, J. & Simmonds, P. (2011). The GB viruses: a review and proposed classification of GBV-A, GBV-C (HGV), and GBV-D in genus Pegivirus within the family Flaviviridae. *Journal of General Virology,* 92(Pt 2): 233-246. ISSN: 1465-2099

Sulbarán, M.Z., Di Lello, F.A., Sulbarán, Y., Cosson, C., Loureiro, C.L., Rangel, H.R., Cantaloube, J.F., Campos, R.H., Moratorio, G., Cristina, J. & Pujol, F.H. (2010). Genetic history of hepatitis C virus in Venezuela: high diversity and long time of evolution of HCV genotype 2. PLoS One. 5(12): e14315. ISSN: 1932-6203

Sulbaran, Y., Gutierrez, C.R., Marquez, B., Rojas, D., Sanchez, D., Navas, J., Rovallo, E. & Pujol, F.H. (2010). Hepatitis A virus genetic diversity in Venezuela: exclusive circulation of subgenotype IA and evidence of quasispecies distribution in the isolates. *Journal of Medical Virology,* 82(11): 1829-1834. ISSN: 1096- 907

Sulkowski, M.S. (2008). Viral hepatitis and HIV coinfection. *Journal of Hepatology,* 48(2): 353-367. ISSN: 0168-8278

Tatematsu, K., Tanaka, Y., Kurbanov, F., Sugauchi, F., Mano, S., Maeshiro, T., Nakayoshi, T., Wakuta, M., Miyakawa, Y. & Mizokami, M., (2009). A genetic variant of hepatitis B virus divergent 1 from known human and ape genotypes 2 isolated from a Japanese patient and provisionally assigned to new genotype J. *Journal of Virology,* 83(20): 10538-10547. ISSN: 1098-5514

Tang, H. & Grisé, H. (2009). Cellular and molecular biology of HCV infection and hepatitis. *Clinical Science*, 117(2): 49-65. ISSN: 0143-5221

Taylor, J.M. (2009). Replication of the hepatitis delta virus RNA genome. *Advances in Virus Research*, 74: 103-121. ISSN: 0065-3527

Tebit, D.M., Nankya, I., Arts, E.J. & Gao, Y. (2007). HIV diversity, recombination and disease progression: how does fitness "fit" into the puzzle? *AIDS Reviews*, 9(2): 75-87. ISSN:1139-6121

Thakur, V., Guptan, R.C., Kazim, S.N., Malhotra, V. & Sarin, S.K. (2002). Profile, spectrum and significance of HBV genotypes in chronic liver disease patients in the Indian subcontinent. *Journal of Gastroenterology and Hepatology*, 17(2): 165-70. ISSN: 1440-1746

Thedja, M.D., Muljono, D.H., Nurainy, N., Sukowati, C.H., Verhoef, J. & Marzuki, S. (2011). Ethnogeographical structure of hepatitis B virus genotype distribution in Indonesia and discovery of a new subgenotype, B9. *Archives of Virology*, 156(5): 855-868. ISSN:0304-8608

Torresi, J. (2002). The virological and clinical significance of mutations in the overlapping envelope and polymerase genes of hepatitis B virus. *Journal of Clinical Virology*, 25(2): 97-106. ISSN: 1386-6532

Toyoda, H., Fukuda, Y., Koyama, Y., Takamatsu, J., Saito, H. & Hayakawa, T. (1997). Effect of immunosuppression on composition of quasispecies population of hepatitis C virus in patients with chronic hepatitis C coinfected with human immunodeficiency virus. *Journal of Hepatology*, 26(5): 975–982. ISSN: 0168-8278

Tran, T.T., Trinh, T.N. & Abe, K., 2008. New complex recombinant genotype of hepatitis B virus identified in Vietnam. *Journal of Virology*, 82(11): 5657-5663. ISSN: 1098-5514

Wang, S.Y., Wu, J.C., Chiang, T.Y., Huang, Y.H., Su, C.W. & Sheen, I.J. (2007). Positive selection of hepatitis delta antigen in chronic hepatitis D patients. *Journal of Virology*, 81(9): 4438-4444. ISSN: 1098-5514

Weiner, A.J., Geysen, H.M., Christopherson, C., Hall, J.E., Mason, T.J., Saracco, G., Bonino, F., Crawford, K., Marion, C.D., Crawford, K.A., Brunetto, M., Barrt, P.J., Miyamura, T., Mchutchinson, J. & Houghton, M. (1992). Evidence for immune selection of hepatitis C virus (HCV) putative envelope glycoprotein. *Proceedings of the National Academy of Sciences* USA, 89(8): 3468-3472. ISSN: 1091- 6490

Wu, J.C., Hsu, S.C., Wang, S.Y., Huang, Y.H., Sheen, I.J., Shih, H.H. & Syu, W.J. (2005). "Defective" mutations of hepatitis D viruses in chronic hepatitis D patients. World Journal of Gastroenterology, 11(11): 1658-1662. ISSN: 1007-9327

Xiang, J., George, S.L., Wunschmann, S., Chnag, Q., Klinzman, D. & Stapleton, J.T. (2004). Inhibition of HIV-1 replication by GB virus C infection through increases in Rantes, MIP-1α, MIP-1β, and SDF-1. *Lancet*, 363(9426): 2040-2046. ISSN: 1474-547X

Yang, H.I., Yeh, S.H., Chen, P.J., Iloeje, U.H., Jen, C.L., Su, J., Wang, L.Y., Lu, S.N., You, S.L., Chen, D.S., Liaw, Y.F., Chen, C.J. & REVEAL-HBV Study Group. (2008). Associations between hepatitis B virus genotype and mutants and the risk of hepatocellular carcinoma. *Journal of the National Cancer Institute*, 100(16): 1134-1143. ISSN: 0027-8874

Yang, W. & Summers, J. (1998). Infection of ducklings with virus particles containing linear double-stranded duck hepatitis B virus DNA: illegitimate replication and reversion. *Journal of Virology*, 72(11): 8710-8717. ISSN: 1098-5514

Yu, M.W., Yeh, S.H., Chen, P.J., Liaw, Y.F., Lin, C.L., Liu, C.J., Shih, W.L., Kao, J.H., Chen, D.S. & Chen, C.J. (2005). Hepatitis B virus genotype and DNA level and hepatocellular carcinoma: a prospective study in men. *Journal of the National Cancer Institute*, 97(4): 265-272. ISSN: 0027-8874

Yuen, M.F., Fung, J., Wong, D.K. & Lai, C.L. (2009). Prevention and management of drug resistance for antihepatitis B treatment. *The Lancet Infectious Diseases*, 9(4): 256-264. ISSN: 1473-3099

Zein, N.N. (2000). Clinical significance of hepatitis C virus genotypes. *Clinical Microbiology Reviews*, 13(2): 223-235. ISSN: 1098-6618

Zeuzem, S., Feinman, S.V., Rasenack, J., Heathcote, E.J., Lai, M.Y., Gane, E., O'Grady, J., Reichen, J., Diago, M., Lin, A., Hoffman, J. & Brunda, M.J. (2000). Peginterferon alfa-2a in patients with chronic hepatitis C. *The New England Journal of Medicine*, 343(23): 1666-1672. ISSN: 1533-4406

Zhu, A.X. & Chung, R.T. (2003). Hepatic steatosis in patients with chronic hepatitis C virus infection. Another risk factor for hepatocellular carcinoma? *Cancer*, 97: 2948-2950. ISSN: 1097-0142

3

Metabolic Aspects of Hepatitis C Virus Infection

Jee-Fu Huang[1,2], Ming-Lung Yu[1,2,3], Chia-Yen Dai[1,2], Wan-Long Chuang[1,2,*] and Wen-Yu Chang[1]
[1]*Hepatobiliary Division, Department of Internal Medicine, Kaohsiung Medical University Hospital*
[2]*Faculty of Internal Medicine, College of Medicine, Kaohsiung Medical University*
[3]*Department of Internal Medicine, Kaohsiung Municipal Ta-Tung Hospital*
Taiwan

1. Introduction

Hepatitis C virus (HCV) infection is one of the most important causes of cirrhosis and hepatocellular carcinoma (HCC) and has a tremendous impact on public health worldwide. It is estimated that there are more than 170 million people chronically infected with HCV, and 3 to 4 million persons are newly infected annually. The risk for developing cirrhosis 20 years after initial HCV infection among those chronically infected varies between studies, but is estimated at around 10%-15% for men and 1-5% for women. Once cirrhosis is established, the rate of developing HCC is at 1%-4% per year. Approximately 280,000 deaths per year are related to HCV infection. HCV-related end-stage liver disease and HCC have become the leading cause for liver transplantation globally.

HCV *per se* is both hepatotropic and lymphotropic. Replication of HCV in diseased extrahepatic organs and tissues may have cytopathic effects. It, therefore, may either trigger latent autoimmunity or induce an autoimmune disease *de novo*. The concise context of pathogenic mechanisms is complex and remains to be elucidated. Generally, in addition to established liver injury, there are multiple examples of extrahepatic metabolic disorders attributed to HCV infection, such as type 2 diabetes mellitus (T2DM), thyroiditis, glomerulopathy, mixed cryoglobulinemia, and other immunological abnormalities. These disorders possess some extent of impact on the disease activity, disease course, clinical outcomes, and treatment efficacy of current therapy.

The aims of this chapter initially reviewed recent studies in terms of the metabolic manifestations of chronic HCV infection (CHC), e.g. proteinuria, T2DM, lipid abnormalities and metabolic syndrome to elucidate the characteristics of metabolic abnormalities and their clinical relevants from the aspect of epidemiological view.

---

* Corresponding Author

Secondly, robust epidemiological data demonstrated the mutual link between HCV infection and T2DM. HCV plays a direct pathogenic role in the emergence of T2DM even in the early stage of liver histological changes and its diabetogenic role has been confirmed. Although the precise mechanisms whereby HCV infection leads to insulin resistance (IR) and glucose abnormalities are not fully clear, it differs from the usual pathogenesis of T2DM in those with non-HCV liver diseases. This chapter also tried to clarify the mutual roles of IR and CHC with respect to the prediction of treatment efficacy, how treatment response affects IR, and the role of pancreatic beta cell function in the interesting suite.

Liver has long been regarded as the key player manipulating complex biochemical metabolism which is essential to maintenance of homeostasis. Recently, studies regarding cytokines have been prevailing worldwide in the past decade. We also introduced several translational studies aiming to elucidate the specific roles of the newcomers in a clinical setting. It will be helpful to clarify the host viral interaction and possible pathogenic mechanisms of this topic.

2. The epidemiological links

2.1 Nephropathy

Viral hepatitis infection *per se* may lead to nephropathy. This uncommon extrahepatic manifestation might be induced either by direct cytopathic effect of virus or interplay between viral, host and environmental factors (Congia et al., 1996; Mazzaro et al., 2000; Strassburg, et al., 1996). The association between hepatitis B virus (HBV) infection and renal involvement, mainly membranous nephropathy, was first reported in 1971(Combes et al., 1971). Deposition of Australia antigen-antibody complexes in glomerular basement membrane was identified in renal tissue of glomerulonephritis. Since then many reports have been made suggesting the association between HBV infection and nephropathy (Johnson et al., 1990; Levo et al., 1977). However, the natural history and the pathogenesis are not well understood but are believed to be mediated by deposition of immune complexes of HBV antigens in the glomeruli (Ito et al., 1981).

On the other side, hepatitis C virus (HCV) has been shown to be a lymphotropic as well as a hepatotropic virus (Mazzaro et al., 2000). Replication of HCV in diseased extrahepatic organs and tissues may have cytopathic effects. It, therefore, may either trigger latent autoimmunity or induce *de novo* an autoimmune disease (Hadziyannis, 1997). HCV-associated nephropathy has been postulated to be a distinct extrahepatic manifestation, which might be related to interplay between intrinsic renal disease, autoimmune abnormality, and host susceptibility (Congia et al., 1996; Garini et al., 2005; Meyers et al., 2003). It occurs largely in the context of cryoglobulinemia (Garini et al., 2005). There are also quite a number of cases who do not have cryoglobulinemia but still present with one of the major HCV- associated nephropathy namely membranoproliferative glomerulonephritis, membranous nephropathy, and focal segmental glomerulosclerosis, the latter two generally not associated with cryoglobulinemia (Meyers et al., 2003). Barsoum et al suggested that in addition to a renal microstructure suitable for sieving out macromolecules associated with HCV infection, such as cryoglobulins and immune complexes, there are several circumstances in which renal cells may be a target for viral cytopathic effects. These effects include ingredients for HCV attachment, endocytosis, and entry. Upregulation of Toll-like receptors and metalloproteinases in mesangial inflammation when there is HCV infection may induce injury through an aggravated immune response. HCV also can enter cells and replicate in B lymphocytes, and it is associated with AA amyloidosis in class VI schistosomal glomerulopathy (Lee et al., 2010). HCV-associated glomerulonephritis, together with interstitial tubular injury, may cause the subsequent CKD.

Previous studies have shown that in persons with HCV infection, higher incidence of microalbuminuria and proteinuria did occur in CHC patients than in those with other forms of liver disease (Liangpunsakul et al., 2005; Muramatsu et al., 2000). Liangpunsakul et al conducted the first nested case-control study to examine the association between microalbuminuria and HCV infection by using the Third National Health and Nutrition Examination Survey (NHANES III) database. The prevalence of microalbuminuria in patients with HCV infection was 12.4%, which was significantly higher than in controls (7.5%) (P= 0.001). This difference persisted even after excluding diabetics from the analysis (11.4% vs. 6.7%) (P= 0.001). In nondiabetic persons with HCV infection, microalbuminuria occurs more often in those who aged 20 years or more (Liangpunsakul et al., 2005).

Proteinuria has been shown to be an early diagnostic marker of kidney damage, and can predict the progression of renal disease in patients with diabetes as well as cardiovascular morbidity and mortality in diabetic and general population (Diercks et al., 2002). During the past decade, proteinuria has taken on a new importance, and is shown to be a cardinal sign and an independent risk factor for the outcome of both kidney and cardiovascular disease (Diercks et al., 2002). Not only is proteinuria associated with glomerular injury and loss of its normal permselective properties, but experimental data have demonstrated that protein-tubular cell interactions have inflammatory and fibrogenic consequences that can contribute to interstitial damage and fibrosis (Keane, 2000).

Two large-scale community studies have been conducted in 2 different HCV-endemic areas in Taiwan. Taiwan is a country prevalent for HBV infection and several scattered hyperendemic areas for HCV infection have been discovered in the past decades. The unique background thus provides a better scope of view to assess the difference of association with nephropathy between HBV and HCV infections. Lee et al recruited 54,966 adults in a county endemic for both HBV (9.9%) and HCV (9.4%) infections. There was a significant increasing trend of HCV seropositivity in parallel with CKD stages, ranging from 8.5% in stage 1 to 14.5% in CKD stages 4-5 (Figure 1) (Lee et al., 2010).

Huang et al also conducted a prospective community-based study enrolling 10,975 subjects to compare the prevalence of proteinuria among adult population in another HBV (13.1%) and HCV (6.5%) endemic area of southern Taiwan (Yang et al., 2006; Yu et al., 2001; Yu et al., 2001; Huang et al., 2006). Setting proteinuria as urine dipstick test ≥1+, the prevalence of proteinuria among subjects seropositive for anti-HCV was significantly higher than HBV carriers and those negative controls. The prevalence of HCV infection was significantly higher among proteinuria than that of non-proteinuria individuals. Multivariate logistic regression analyses showed HCV infection was an independent significant factor associated with proteinuria. The significance remained even excluding those diabetics. By contrast, HBV infection did not show significant difference between proteinuria and non-proteinuria subjects. The robust epidemiological data demonstrated the consistent and significant association between HCV infection with both CKD prevalence and CKD disease severity.

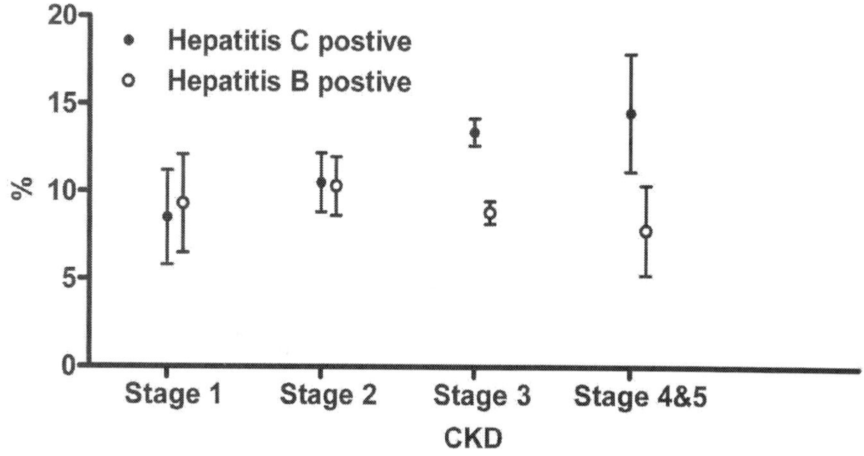

Fig. 1. Prevalence of hepatitis C virus infection and hepatitis B virus infection based on the staging of chronic kidney disease (CKD). Error bars represent 95% confidence intervals.

2.2 Type 2 diabetes mellitus

HCV infection and Type 2 diabetes mellitus (T2DM) are two major rising epidemics haboring tough challenges both to the clinicians and to the health-care systems in terms of diagnostic, therapeutic, and economic implications (Aytaman et al., 2008). In 2003, there were over 194 million people diagnosed with diabetes worldwide, with an estimated 333 million persons to be afflicted by 2025. These numbers, however, do not include those with undiagnosed diabetes, a population that is currently estimated to represent over 26% of the U.S. population alone. On the other side, HCV infection currently affects 3% of the world population, estimated 170 million people worldwide, leading to a substantial rise in the prevalence of chronic liver disease, with its enormous health and economic impacts. Although type 1 DM has been observed in patients treated with interferon (IFN), the majority of HCV-related diabetes is T2DM.

T2DM is a common endocrine disorder encompassing multifactorial pathogenetic mechanisms. These mechanisms include resistance to the action of insulin, increased hepatic glucose production, and a defect in insulin secretion, all of which contribute to the development of overt hyperglycemia (Saltiel 2001). Although the precise mechanisms whereby these factors interact to produce glucose abnormalities are uncertain, it has been suggested that the final common pathway responsible for the development of T2DM is the failure of the pancreatic beta-cells to compensate for the insulin resistance (IR). The biological mechanism underlying IR or T2DM in HCV infection remains unclear. Shintani et al demonstrated that the ability of insulin to lower the plasma glucose level was impaired without gain in body weight at young age in HCV core gene transgenic mice study. A high level of tumor necrosis factor-alpha was considered to be one of the bases of IR, which act by disturbing tyrosine phosphorylation of insulin receptor substrate (IRS)-1, a central molecule of the insulin-signaling cascade. These findings provided a direct experimental evidence for the contribution of HCV in the development of IR and in the pathogenesis of T2DM (Shintani et al., 2004). In addition, clinical study by Kawaguchi et al indicated an increase in fasting insulin levels was associated with the presence of serum HCV core protein, the severity of hepatic fibrosis, and a decrease in expression of IRS-1 and IRS-2 in patients with HCV infection. More severe IR was present in noncirrhotic patients with HCV infection than in patients with other liver diseases (Kawaguchi et al., 2004). HCV core-induced suppressor of cytokine signalling 3 may promote proteosomal degradation of IRS1 and IRS2 through ubiquitination, and which may be a unique mechanism of HCV-associated IR. In patients with undetectable levels of HCV core protein, fasting insulin levels were within the normal range. In contrast, in patients with detectable levels of HCV core protein, fasting insulin levels were increased. Thus, HCV core protein seemed to play a crucial role in HCV-associated IR.

There are strong evidence supporting an epidemiologic link between CHC and T2DM(Allison et al., 1994; Mason et al., 1999; Mehta et al., 2000; Wang et al., 2003; Zein et al., 2005). The association between T2DM and CHC was first reported by Allison et al, who observed that the prevalence of T2DM was significantly higher in those with HCV-related cirrhosis than those with cirrhosis resulting from other liver diseases (Allison et al., 1994). The diagnosis of HCV infection and the identification of risk factors for HCV infection preceded the diagnosis and/or onset of T2DM in anti-HCV(+) diabetics (Knobler et al., 2000). The prevalence of T2DM and impaired fasting glucose (IFG) was higher among HCV-infected patients with advanced versus those with early histological disease. Advanced histological disease predicted T2DM/IFG after controlling for other identified risk factors for T2DM (Zein et al., 2005). Similar features were also observed from Asia Pacific region. Huang et al, in a community survey composing serological and virological features of HCV infection, further extended the observation that HCV viremia, but not anti-HCV seropositivity alone, increased the association with T2DM (Figure 2). It may imply that a persistent and/or active phase of HCV infection is associated with T2DM (Huang et al., 2007). Wang et al prospectively followed 4,958 persons aged 40 years or more without T2DM from a community-wide cohort in southern Taiwan for 7 years. The 7-year cumulative incidence of those anti-HCV+ subjects was nearly 2-folds increase than those HBsAg+ and negative controls. After stratification by age and body mass index (BMI), the risk ratio for T2DM in anti-HCV+ participants increased when age decreased and BMI levels increased (Wang et al., 2007). Generally, the prevalence of anti-HCV seropositivity in the T2DM population ranged from 1.8 to 12.1%, whereas T2DM developed in 14.5 to 33.0% of CHC patients (Allison et al., 1994; Antonelli et al., 2005; Caronia et al., 1999; Lecube et al., 2006; Mason et al., 1999; Mehta et al., 2003; Mehta et al., 2000; Zein et al., 2005). Different background in terms of ethnicity, age, prevalence of T2DM, BMI, viral load and genotype might contribute to the divergent results of the epidemiological observations. With respect to epidemiological aspect, HCV is considered to be diabetogenic and T2DM represented one more disease to be included in the list of established extrahepatic manifestations of HCV infection.

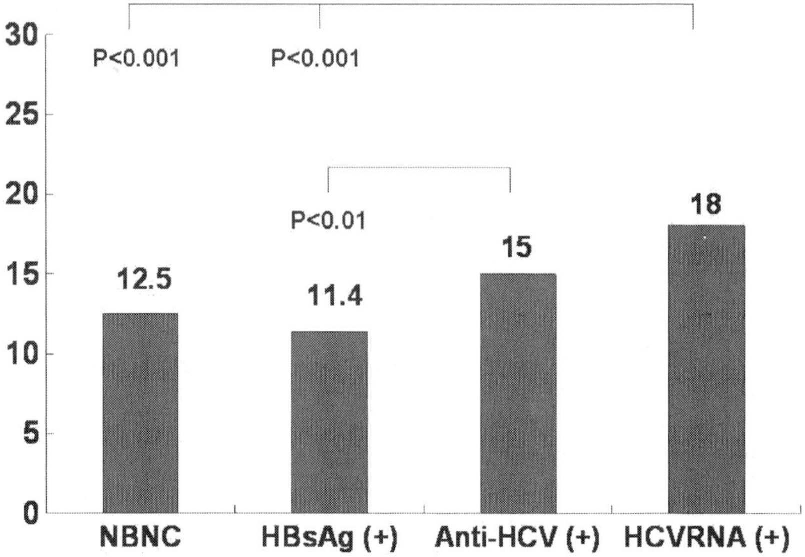

Fig. 2. Prevalence of T2DM among subjects with different etiologies of viral hepatitis. HBsAg: hepatitis B surface antigen; anti-HCV: hepatitis C virus antibodies; NBNC: neither HBsAg (+) nor anti-HCV (+); HCVRNA: hepatitis C virus RNA.

The relationship between T2DM and HCV genotypes remains controversial (Knobler et al., 2000; Lecube et al., 2006; Mason et al., 1999; Petit et al., 2001). Zignego et al demonstrated that HCV genotype-2a (G-2a) was specifically linked with extrahepatic manifestations such as cryoglobulinemia (Zignego et al., 1996). An association between G-2a infection and T2DM was also reported (Mason et al., 1999). However, no association was found between fasting insulin levels and HCV genotypes in Japanese study (Kawaguchi et al., 2004). Taiwanese study showed that neither HCV G-1 nor G-2 infection was significantly associated with T2DM (Huang et al., 2007). Recently a large-scale international collaborative study addressing the association between IR and viral clearance in G-1, -2 and -3 patients showed that IR was more common in patients with G-1 than in those with G-2/3 infection. Viral eradication was associated with a reduction in IR in patients infected with G-1 but not in those with G-2/3 infection, suggesting a causal relationship between G-1 infection and IR in vivo (Thompson et al., 2011).

2.3 Metabolic disorders

Although the precise mechanisms leading to HCV cell entry are not fully elucidated, HCV uptake by hepatocytes has been demonstrated to be mediated by the low-density lipoprotein (LDL) receptor as well as other proteins such as tetraspanin CD81, claudin-1 and occludin. Experiments in vitro showed competitive inhibition of binding between HCV and LDL-receptor by LDL-C (Monazahian et al., 1999). HCV cell entry is also mediated by another lipoprotein receptor, the scavenger receptor class B type I, which is responsible for the entry of various classes of lipoproteins, primarily the high-density lipoproteins (HDL) (Negro, 2010). In addition, lipoproteins play an important role in the process of HCV infection since complexing of the virus to the low-density lipoprotein (VLDL-C) or LDL-C could promote endocytosis of HCV via the LDL receptor (Agnello et al., 1999). HCV infection and hypobetalipoproteinemia is characteristic in patients with HCV infection (Monazahian et al., 1999; Monazahian et al., 2000; Perlemuter et al., 2002). There are a lower cholesterol (total cholesterol, HDL-C and LDL-C) and a lower triglycerides (TG) levels in CHC patients than those of normal subjects (Dai et al., 2008; Siagris et al., 2006). Dai et al conducted a community-based mass-screening of 11,239 adults showing the subjects with normal serum cholesterol and TG levels had significantly higher proportion of anti-HCV+ than those who had elevated serum cholesterol and/or elevated TG levels. For anti-HCV+ patients, patients with normal serum cholesterol and TG levels had significantly higher proportion of positive HCV RNA than those who had elevated serum cholesterol and/or elevated TG levels. It is particularly noteworthy that subjects with HCV viremia have significantly lower serum cholesterol and TG levels than those who were negative for HCV RNA (Dai et al., 2008). Furthermore, a HCV genotype-based different impact on the lipid profile and hepatic steatosis was reported as well (Moriya et al., 2003; Serfaty et al., 2001). These epidemiological findings imply that HCV plays a significant role on serum lipid profiles and lipids are essential to the HCV life cycle. Therefore, alteration of blood lipids is a characteristic feature of HCV infection clinically. Recent studies demonstrated that successful eradication of HCV infection was associated with a significant increase in total cholesterol, LDL-C and TG levels in G-1, -2, and -3 infection (Tada et al., 2009). This increase from baseline was most pronounced in patients with G-3 infection(Thompson et al., 2011).

Metabolic syndrome (MS) is a complicated disorder encompassing clinical features of obesity, hyperglycemia, hypertension, dyslipidemia and IR. It carries a high risk for future development of micro- and macrovascular complications. Atherosclerosis and T2DM, as major subsequent events of MS, are critical health issues globally (Grundy et al., 2005). There is groundswell evidence indicating that the atherosclerotic process is regulated by intervening inflammatory mechanisms. IR, heroine of the scenario in the pathogenesis of MS, has been increasingly recognized as playing a key role in the inflammatory processes. Histologically, hepatic steatosis is a common feature of CHC, which is observed in 30-70% of the patients (Hsieh et al., 2007; Watanabe et al., 2008). Many factors are known to be risks for hepatic steatosis, including DM, hyperlipidemia, and obesity (Sanyal, 2005). Apart from its hepatotropic characteristic, HCV infection carries a significant pathogenic effect for development of IR, albeit the underlying biological mechanisms are diverse and multifactorial. Metabolic abnormalities including liver steatosis, obesity and DM can also worsen the course of CHC (Moucari et al., 2007; Tarantino et al., 2006).

The prevalence of the MS in patients with CHC varied from 4.4% in Italy, 24.7% in Taiwan and 51% in US veterans (Huang et al., 2009a; Keane et al., 2009; Svegliati-Baroni et al., 2007), suggesting that host factors such as ethnic origin could play a role in the association between IR and CHC (Conjeevaram et al., 2007; Serste et al., 2010). Liangpunsakul et al examined the relationship between nondiabetic subjects with HCV infection and microalbuminuria by using NHANES III database which consisted of 15,336 adults from the United States. There was no difference in the MS prevalence between HCV group and controls (Liangpunsakul et al., 2005). However, subjects with HCV infection carrying a higher prevalence of MS than controls was observed from an HCV-hyperendemic area (>35% of anti-HCV+ prevalence in adults) in Taiwan. Those anti-HCV+ residents had a higher waist circumference and a higher prevalence of hypertension as the common features of MS (Huang et al., 2009). Further exploration is needed to clarify the complex context of MS with respect to different ethnicity. Furthermore, emerging lines of clinical data revealed that several metabolic disturbance; such as obesity, IR, and hepatic steatosis, are significant risk factors for decreased treatment response to pegylated IFN (PegIFN) and ribavirin combination therapy in CHC patients (Bressler et al., 2003; Camma et al., 2006; Romero-Gomez et al., 2005; Watanabe et al., 2008). The association between HCV infection and MS could be validated by the sequential changes of MS characteristics upon treatment response to antiviral therapy. In addition, the difference of atherosclerotic cardiovascular disease risk between MS subjects with HCV infection and non-HCV subjects deserves to be further elucidated.

3. Hepatitis C virus infection and type 2 diabetes mellitus

3.1 The characteristics of glucose abnormalities in patients with chronic hepatitis C infection

There has been emerging and robust data, particularly in the aspect of epidemiology, suggesting that a higher prevalence of T2DM among patients with HCV infection, whereas HCV infection is a risk factor for developing T2DM.

T2DM is often present at least 4 to 7 years before diagnosis (Harris et al., 1992). Therefore, definitive diagnosis of glucose abnormalities is an important issue because it allows attempts to improve clinical outcomes, such as weight reduction and lifestyle modification (Gerstein et al., 2006; Tuomilehto et al., 2001). On the other side, almost all T2DM patients have experienced the prediabetic condition, namely, IFG and/or impaired glucose tolerance (IGT), before a definite diagnosis of T2DM was made. In addition to future DM development, the prediabetic condition also carries a risk for cardiovascular disease (Kannel et al., 1979). Generally and commonly, fasting plasma glucose (FPG) level alone is used as a screening test for the diagnosis of DM. However, this practice is based on the relative convenience and lower cost of FPG compared with a 75-gram oral glucose tolerance test (OGTT) (Grundy et al., 2005; Knowler et al., 2002). The discrepancy of distribution of glucose abnormalities before and after OGTT was estimated that 19.3–59.3% of glucose abnormalities remained undetected using the current IFG criteria alone (Huang et al., 1999; 2003).

The same scenario exists in the link between glucose abnormalities and HCV infection. Previous data linking HCV infection and DM mainly focused on patients with overt DM. Lecube et al prospectively recruited a total of 642 hispanic patients (498 anti-HCV+ and 144 anti-HCV-) in a single center. Patients were classified as having chronic hepatitis (n = 472) or cirrhosis (n = 170) by means of clinical manifestations or liver biopsy. A 3 folds increase in the prevalence of glucose abnormalities was observed in HCV+ patients with chronic hepatitis in comparison with HCV- subjects (32% vs. 12%). In addition, 18% more new DM cases and 30% more new cases of IGT were uncovered by OGTT in anti-HCV+ patients, which were significantly higher than those values in anti-HCV- patients (Lecube et al., 2004). Another case-control study recruited 683 CHC patients and 515 sex-/ age-matched community-based controls were conducted in Taiwan aiming to elucidate the entire suite of glucose abnormalities. OGTT was performed in 522 CHC patients and 447 controls without known T2DM. The prevalence of normoglycemia, IGT, and T2DM in 683 CHC patients was 27.7%, 34.6%, and 37.8%, respectively. Of note is 18.6% of CHC patients who readily met with DM criteria were undiagnosed (Figure 3). For those without known DM, there were 3.5 folds increase in the prevalence of glucose abnormalities in CHC patients in comparison with controls (Huang et al., 2008) (Table 1). The two studies implied that CHC patients carried a high prevalence of glucose abnormalities and also suggested that determination of glucose abnormalities by OGTT should be indicated in CHC patients. It might also suggest that different criteria are necessary for DM diagnosis in patients with HCV infection, such as a lower cut-off for normoglycemia, prediabetes, and DM.

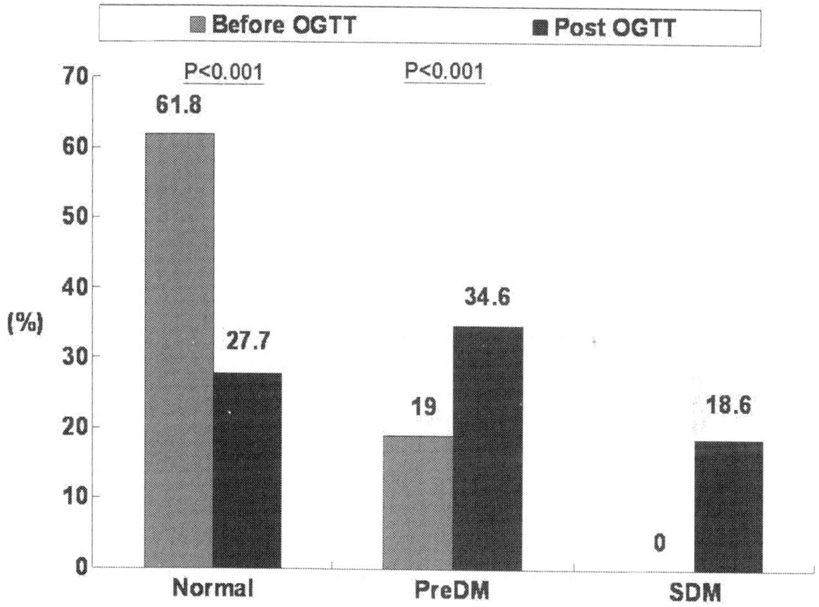

Fig. 3. Distribution of glucose abnormalities among CHC patients before and after OGTT. OGTT: 75-g oral glucose tolerance test; PreDM: prediabetes, SDM; subclinical diabetes.

The cause-effect interaction between a common endocrine disorder and an infectious disease is an important issue to elucidate. Comparison between the stages of glucose abnormalities and disease severity of HCV infection may pave a way to the clarification. HCV core gene transgenic mice study by Shintani et al demonstrated that the ability of insulin to lower the plasma glucose level was impaired without gain in body weight at young age (Shintani et al., 2004). The presence of advanced histological disease in genetically predisposed CHC patients was associated with a higher prevalence of DM and IFG (Zein et al., 2005). There was also a significant linear trend from normoglycemia to T2DM in terms of age, family history of T2DM, and advanced liver fibrosis in CHC patients (Huang et al., 2008). Both experimental and clinical studies provide the evidence that the genuine connection between

HCV infection and T2DM is initiated at early stages of liver disease. HCV infection may contribute to the subtle development of glucose abnormalities at young age.

Characteristics	Anti-HCV (-) Controls	Anti-HCV (+) patients	P value	OR, 95% CI
Patients, *n*	447	552		
Age (years)	50.5 ± 13.7	52.0±12.4	NS	
Male, *n* (%)	222 (49.7)	268 (48.6)	NS	
Normoglycemia, n (%)	289 (64.7)	189 (34.2)	<0.001	0.29, 0.22-0.37
Prediabetes, n (%)	145 (32.4)	236 (42.8)	0.001	1.56, 1.20-2.02
Subclinical DM, n (%)	13 (2.9)	127 (23.0)	<0.001	9.98, 5.55-17.9

Table 1. Characteristics of CHC patients without known DM and controls and their glucose abnormalities validated by OGTT. OR: Odds ratio; CI: confidence interval

3.2 Insulin resistance predicts response to peginterferon-alpha/ribavirin combination therapy in chronic hepatitis C patients

Liver has long been regarded as the key player manipulating the homeostasis of glucose metabolism. As a largest reservoir of glucose, the importance of liver in the glucose metabolism draws its much attention from the patients with advanced liver disease. Hepatic diabetes was once recognized when diabetes developed in those who had advanced liver cirrhosis or severe liver injury, on which overt fasting hypoglycemia and/or postprandial hyperglycemia appeared as a common phenomenon. IR is therefore a common feature of some liver diseases, especially with advanced stages. To address the relationship between IR and liver diseases, the extent of liver injury and the impact of liver disease itself on insulin signalling and subsequent glucose metabolic dysregulation should be taken into consideration. Besides skeletal muscle and adipose tissue, liver is the major targets for the metabolic actions of insulin. Insulin regulates glucose homeostasis by reducing hepatic glucose output and by increasing the rate of glucose uptake by skeletal muscle and adipose tissue.

The measurement of IR in the field of hepatology remains to be investigated in a clinical setting. Although the "gold standard" test for IR assessment is the hyperinsulinemic-euglycemic clamp test. However, the difficulty of techniques and its laborious characteristic much embarrass the wide use of the test. The Homeostatic Model Assessment (HOMA), which has been used in large epidemiological studies, offers an estimate of IR by multiplying the FPG and insulin concentrations and dividing this product by 22.4 or 403 when the glucose concentration is expressed as millimolar or milligrams per deciliter, respectively. A HOMA score close to 1 indicates normal insulin sensitivity, whilst an increase IR circumstance is recognized as the score over 2. HOMA has the advantage of requiring only a single fasting plasma sample measured for glucose and insulin. Nonetheless, the lack of standardization of insulin assays and the difference of IR standardization between races may undermine the accuracy of the measurements (Neuschwander-Tetri, 2008).

HCV may induce IR irrespective of the severity of liver disease and IR may be associated with severe hepatic fibrosis and contribute to fibrotic progression in CHC (Alexander, 2000; Dai et al., 2008; Hui et al., 2003; Petit et al., 2001; Taura et al., 2006). There was also a dose-response relationship between HCV RNA level and the presence of IR, whilst IR was positively associated with the severity of hepatic steatosis (Hsu et al., 2008).

Combination therapy with PegIFN and ribavirin has been recommended as standard therapy for patients with HCV infection with favorable efficacy (Hui et al., 2003; Petit et al., 2001; Svegliati-Baroni et al., 2007). It therefore provides a wide scope of view addressing the correlation between HCV infection and IR. Several clinical predictors of the sustained virologic response (SVR), namely, HCV RNA negativity 24 weeks after treatment, to combination therapy have been elucidated such as the viral factors (e.g. viral G-2 or -3, lower pretreatment viral load) (Hui et al., 2003; Petit et al., 2001; Strader et al., 2004; Svegliati-Baroni et al., 2007; Taura et al., 2006) and host factors (younger age, lower BMI, non-African-American or Asian races and interleukin-28B polymorphisms, etc.)(Fried et al., 2002; Strader et al., 2004; Svegliati-Baroni et al., 2007; Yu et al., 2009; Yu et al., 2011). Glucose abnormalities have also been suggested recently to be a risk factor for nonresponse (Muiret al., 2004). Romero-Gomez et al performed the first study aiming to elucidate impact of IR on the treatment response to PegIFN and ribavirin combination therapy. Achievement of SVR was significantly associated with IR. The SVR rate of G-1 patients with IR (HOMA-IR > 2) was 32.8%, which was significantly lower than that (60.5%) without IR (Romero-Gomez et al., 2005). Dai et al recruited 330 treatment-naïve CHC Taiwanese patients without overt diabetes validated by OGTT. Patients with high HOMA-IR achieved significantly lower rate of SVR than those who with low IR. The significantly lower SVR rate in high HOMA-IR patients than in low HOMA-IR patients was observed in G-1 patients but not in non-G-1 patients (Figure 4). They demonstrated that HCV G-1, pretreatment HCV RNA level and pretreatment HOMA-IR were independent factors associated with SVR. Of note was that IR was associated with SVR, especially among 'difficult-to-treat' patients, i.e., the patients with G-1 infection and high pretreatment viral loads (>400,000IU/mL) (Dai et al., 2009). Conjeevaram et al also showed that IR was independently associated with a lower SVR rate (Conjeevaram et al., 2007). It is noteworthy that the mean HOMA-IR of African and Caucasian American with different levels of steatosis was 3.5 to 6.8, which seemed to be higher than 2.2 of Taiwanese patients. To develop individualized or personalized therapy for CHC, elucidating the changeable predictors of SVR and further manipulating them seems potentially achievable in addition to adjusting the regimens according to the unchangeable viral factors such as HCV genotype. Since IR is considered as a factor which can be modified and improved by various interventions, further prospective studies will be valuable to evaluate whether the effective approaches to improve IR before initiation of the combination therapy for CHC can significantly increase the SVR rate. Taken together, these findings suggest pretreatment HOMA-IR is a predictor for the treatment outcomes of combination therapy.

Previous studies have shown that around one to two thirds of liver biopsies from CHC patients have histological evidence of steatosis (Dai et al., 2006; Poustchi et al., 2008; Romero-Gomez et al., 2005). Hepatic steatosis was associated with overweight, hepatic fibrosis and a high TG level. There are also associations between IR, steatosis and liver fibrosis in CHC patient (Conjeevaram et al., 2007; D'Souza et al., 2005; Hsieh et al., 2007;

Powell et al., 2005). Steatosis and fibrosis also predict for treatment response to PegIFN/ribavirin therapy (Heish et al., 2008; Muzzi et al., 2005). Previous study showed IR but not steatosis was independently associated with lower SVR rate (Conjeevaram et al., 2007; Dai et al., 2009; Romero-Gomez et al., 2005). IR has been suggested as the cause, rather than the consequence, of hepatic steatosis and fibrosis in patients with HCV, particularly those with G-1 infection (Sud et al., 2004). The mechanisms for more obvious and important influence of IR than steatosis and fibrosis might need further studies in the future.

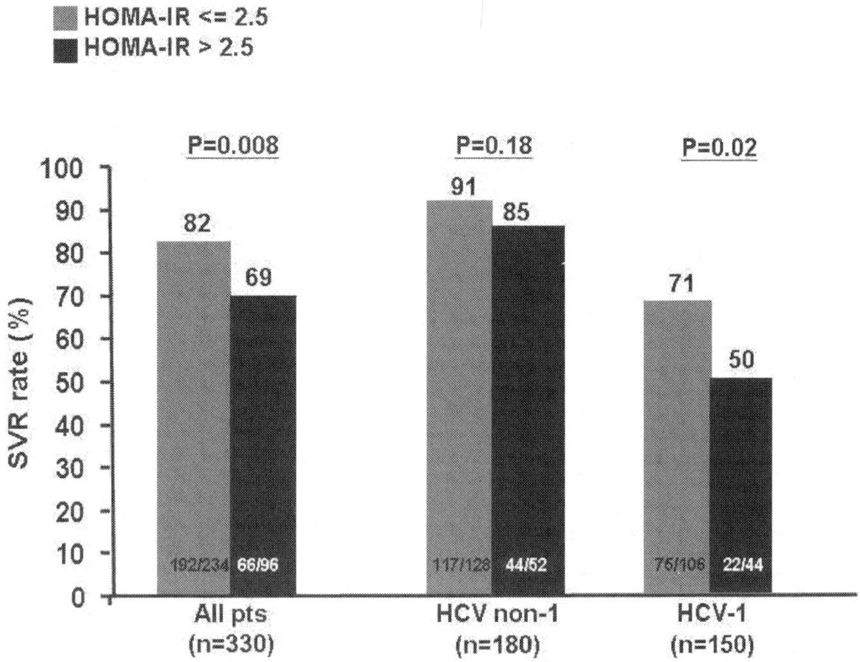

Fig. 4. Sustained virological response rates to combination therapy with pegylated interferon-alpha and ribavirin among chronic hepatitis C patients and stratified by HCV genotype 1b and non-1b infection. The HOMA-IR was defined as high (>2.5 black bars) and low (<= 2.5 white bars). SVR: sustained virological response; HOMA: the homeostasis model assessment; IR, insulin resistance

3.3 The impact of pegylated interferon plus ribavirin therapy on insulin resistance and beta-cell function in chronic hepatitis C patients

PegIFN/ribavirin combination therapy is the current standard of care for the treatment of CHC during the past decade (Fried et al., 2002; Hsieh et al., 2007; Yang et al., 2006; Yu et al., 2008). However, IFN is an integral player in immunity and may exacerbate an existing autoimmune tendency, which may subsequently precipitate immune-mediated abnormalities *de novo* (Kawaguchi et al., 2007). Emergences of IR and subsequent DM have been demonstrated with IFN-based therapy, although the mechanism remains to be clarified

(Borg et al., 2007; Chedin et al., 1996). Therefore, the interplay between IFN-based antiviral therapy and alteration of insulin sensitivity deserved to be elucidated. Several studies have suggested an association between viral clearance and/or suppression and relief of IR by IFN-based therapy. Reduced IR and subsequent improved glucose control after IFN therapy had been shown among patients with CHB and CHC infections (Tai et al., 2003). HOMA-IR decreased after treatment in those responders, whilst it remained unchanged in those non-responders. Kawaguchi et al further demonstrated that clearance of HCV improves IR, beta-cell function, and hepatic IRS1/2 expression by immunostaing, whilst there were no significant changes in IR and beta-cell function after antiviral therapy in those nonresponders and relapsers (Kawaguchi et al., 2007). Recently, results of the HALT-C study indicated that on-treatment virological suppression correlated with reduction in HOMA-IR at week 24. Huang et al further extended the observation in G-1 and -2 patients showing that there was no significant decline of HOMA-IR even in those responders. The significant decline of HOMA-IR after treatment was observed only in those patients with high pretreatment HOMA-IR, irrespective of SVR achievement (Huang et al., 2011). Recent study in a clinical trial cohort of CHC patients showed that SVR was independently associated with a reduction in IR in G-1 but not G-2/3 patients (Huang et al., 2011). The results suggest a causal relationship between specific genotype and IR. The somewhat discordant results may imply that the HOMA-IR with respect to SVR may have been influenced by variables such as race, age, genotypes, validation methods for diabetes, cut-off value of IR, treatment adherence, and/or the presence of liver steatosis. Since the mechanisms involved in the emergence of IR are multifarious, further long-term follow-up study is needed to elucidate in this context.

In addition to hyperinsulinemia, pancreatic beta-cell hyperfunction aiming to maintain glucose homeostasis and elevated serum insulin level is the main feature of glucose abnormalities. The scene is also common in HCV infection, and insulin secretion is increased in the initial stages of HCV infection to compensate for IR development in both experimental and human studies. The formulas for the HOMA model of pancreatic beta-cell function are as follows: HOMA-%B= fasting insulin level (μU/mL) × 360/[FPG (mg/dL) - 63]; the normal value for HOMA-%B was >80% (Matthews et al., 1985). Huang et al recruited 277 non-diabetic Taiwanese CHC patients with adequate treatment adherence and showed there was a significance relief of beta-cell function in CHC patients after PegIFN/ribavirin combination therapy, particularly in those responders. Parallel the results that there was no significant decline of HOMA-IR in those responders, the sequential change of beta-cell function might suggest that beta-cell function was recovered earlier than that of IR in CHC patients receiving PegIFN/ribavirin combination therapy (Huang et al., 2011).

4. Related biomarkers

The National Institutes of Health defined biomarker as a "characteristic that is objectively measured and evaluated as an indicator of normal biological processes, pathogenic processes, or pharmacologic responses to a therapeutic intervention." This broad definition includes information that can be derived from characterizing an individual's genome, transcriptome, proteome, metabolome, markers of subclinical disease, and metabolic end products. Several aspects of effort have been performed dedicating to screen for possible surrogate biomarkers in terms of metabolic manifestations of CHC.

The precise biological mechanisms whereby HCV infection leads to metabolic abnormalities are not fully clear. HCV may induce a Th1 lymphocytes immune-mediated response which leads to activation of tumor necrosis factor (TNF)-α system and elevation of interleukin-6 levels. Meanwhile, HCV directly causes liver steatosis. All the above events may precipitate to the development of liver fibrosis. TNF-α system activation, liver steatosis, and fibrosis contribute to the development of IR, which plays a pivotal role in the development of subsequent metabolic events (Lecube et al., 2006). HCV-induced inflammatory changes may subsequently lead to increased oxidative stress and increased peroxidation, which evoke systemic inflammatory responses (SIR) than other liver diseases (Lecube et al., 2006b). Therefore, SIR triggered by HCV and/or its subsequent immune cascades and cytokine storms may play a major role in the related pathogenic mechanisms in terms of liver injury and the unique extrahepatic manifestations (Figure 5). Meanwhile, SIR may also contribute either directly or indirectly to the disease course, viral response, disease severity, and response to antiviral treatment. Cytokines triggering, which interacts with innate and/or adaptive immune responses, is one of the major concealed players of the scenario.

4.1 Retinol-binding protein 4

For the past decades, several biomarkers have been studied as surrogates of IR. One example is the discovery of retinol-binding protein 4 (RBP4) as a biomarker of IR by DNA microarrays (Tamori et al., 2006; Yang et al., 2005). RBP4, a circulating protein that was highly expressed in the adipose tissue of the adipocyte-specific glucose transporter 4 knockout mice, was demonstrated to be closely related to IR (Yang et al., 2005). IR was induced in mice that were either overexpressing RBP4 or were injected with recombinant RBP4, whereas RBP4 knockout mice showed increased insulin sensitivity. The elevated levels of RBP4 in both adipose tissue and serum were ameliorated when treated with insulin-sensitizer. Reducing serum RBP4 levels ameliorated IR in mice fed a high-fat diet. Graham et al extended this research to humans and showed a correlation between RBP4 levels and the magnitude of IR in subjects with obesity, IGT, or T2DM (Graham et al., 2006). Serum RBP4 levels were even increased in healthy adults with a strong family history of DM. All these interesting findings indicate that RBP4 may serve as a new marker for IR. Lowering RBP4 could be a new strategy for treating T2DM was also deduced then. However, inconsistent observations have been postulated simultaneously and continuously since the discovery of its role in IR (Erikstrup et al., 2006; Janke et al., 2006; Takashima et al., 2006; von Eynatten et al., 2007). Although there was an association between RBP4 and steatosis in G-1 CHC patients, Petta et al suggested RBP4 might be the expression of a virus-linked pathway to steatosis unrelated to IR (Petta et al., 2008).

Liver is the primary source of RBP4 synthesis (80%), the extent of liver injury therefore should be considered when evaluating the correlation of RBP4 with IR (Newcomer and Ong, 2000; Yagmur et al., 2007). Huang et al demonstrated that the increasing change of RBP level from normoglycemia to IGT and T2DM in healthy subjects was indistinct in CHC patients. No correlation of RBP4 levels with HOMA-IR in terms of different stages of glucose tolerance in CHC patients was observed. Intriguingly, in contrast with the parallel increment of IR dependent of histological grading and staging, RBP4 level was inversely correlated with both hepatic necroinflammatory activity and fibrosis stages (Huang et al.,

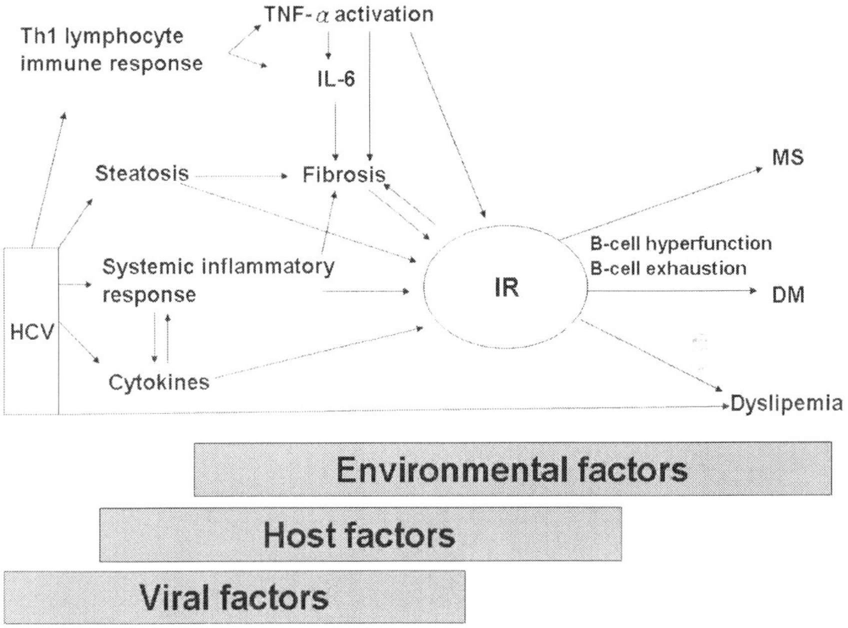

Fig. 5. The possible pathogenic mechanisms leading to the development of insulin resistance and subsequent metabolic disorders. HCV triggers an immune cascades mainly mediated by Th1 lymphocytes. These lymphocytes increase the activation of TNF-α and elevation of interleukin-6 levels. HCV directly leads to steatosis, particularly in those with genotype-3 infection. Meanwhile, HCV may also induce systemic inflammatory response and cytokine storms, which are potentially fibrogenic factors. All the events increase the risk for IR. Fibrosis may be exacerbates by the development of IR, partly by the activation of hepatic stellate cell. IR plays a pivotal role in the development of subsequent metabolic events. With respect to the development of diabetes mellitus (DM), pancreatic beta-cell hyperfunction aiming to maintain glucose homeostasis and elevated serum insulin level is the main feature before overt DM occurs. The scene of pancreatic beta-cell hyperfunction and beta-cell exhaustion may develop in the initial stages of HCV infection. The milieu of host factors (genetic predisposition, male, race, body mass index, etc), environmental factors (sedentary life, diet, etc) and viral factors (genotype, viral load) is also involved into the complex context. TNF-α: tumor necrosis factor-α; IL-6: interleukin-6; IR: insulin resistance; MS: metabolic syndrome; DM: diabetes mellitus

2009). Iwasa et al also postulated the similar results in 81 G-1 CHC patients. Moreover, they observed that only patients who had achieved SVR had higher post-treatment RBP4 levels after PegIFN/ribavirin therapy (Iwasa et al., 2009). These clinical studies suggested that the relationship between RBP4 and IR in general population is diminished in CHC patients because of the negative correlation between RBP4 and disease severity. Therefore, it implied that RBP4 may play a limited role in the identification of IR in patients with HCV infection.

4.2 High-sensitivity C-reactive protein

Acute phase reactions with elicited acute phase proteins directly or indirectly from liver are common features in patients with different extent of liver insults ranging from acute liver injury to advanced liver cirrhosis (Ramadori et al., 2001). The features of acute phase proteins include increased C1-inhibitor, C9, C4 and orosomucoid, whilst decreased transferrin and fetuin/a2HS-glycoprotein in CHC patients who responded to IFN-based antiviral therapy (Biro et al., 2000). In the context of cytokines, high-sensitivity C-reactive protein (hs-CRP) has been shown to be closely related to the occurrence of SIR (Koenig et al., 2008). It plays as a major role in the scenario of various chronic liver diseases, such as non-alcoholic fatty liver disease and nonalcoholic steatohepatitis (Kogiso et al., 2009; Riquelme et al., 2009). Serum hs-CRP levels were elevated in patients with IR and were correlated with MS (Ridker et al., 2003; Yudkin et al., 1999). Moreover, hs-CRP has been shown to predict future risks for cardiovascular disease and related mortality (Zacho et al., 2008; Zethelius et al., 2008). With respect to hepatological views, hs-CRP is a liver specific acute-phase protein, and its expression in hepatocytes is closely related with proinflammatory cytokines such as TNF-alpha, interleukin-1 and interleukin-6 (Kushner, 1982). The relationships between HCV infection and vascular atherosclerosis remain an argument of debate(Ishizaka et al., 2002; Moritani et al., 2005; Oyake et al., 2008). There were discrepant results of the association of hs-CRP level and anti-HCV seropositivity in different study groups (Floris-Moore et al., 2007; Kalabay et al., 2004; Nascimento et al., 2005; Reingold et al., 2008; Tsui et al., 2009; Yelken et al., 2009). Recent studies demonstrated CHC patients had a higher serum hs-CRP level (> 3 mg/L, a condition indicative of a high cardiovascular risk) than healthy controls (Zacho et al., 2008). CHC patients with elevated hs-CRP had substantially higher levels of lipid profiles. However, advanced liver diseases may be associated with decreasing cholesterol level and the extent of liver injury should be taken into account when addressing this issue. Moreover, the lipid profiles are reflected by multiple factors such as race, age, gender, life style and meal habits. Therefore, it might be too conclusive to imply that lipid profiles were the major factors correlated with hs-CRP level in CHC patients. Nevertheless, the observation suggested that hs-CRP may be used as a complementary surrogate marker for cardiovascular risks in CHC patients. Further study addressing the sequential changes and the correlation between lipid profiles and hs-CRP in a long-term follow-up basis will be needed to elucidate this intriguing issue.

Hs-CRP level was significantly decreased after PegIFN/ribavirin combination therapy, particularly for those with viral suppression. It may suggest that SIR in HCV infection could effectively be relieved after antiviral therapy, particularly in patients achieving an SVR. Intriguingly, among those non-SVR patients, decreases of serum hs-CRP levels were significantly observed only in relapsers but not in non-responders. It has been shown that those relapsers have more favorable outcomes after re-treatment programs compared with non-responders (Keeffe, 2005). The decreased SIR in relapsers, which was reflected by significantly decreased hs-CRP level, may in a part contribute to the favorable outcomes. It is noteworthy whether the non-responders carry a higher risk of developing cardiovascular events and it awaits further intervention.

4.3 PI3K/Akt

The precise mechanisms whereby SIR triggered by HCV infection is not fully clear. Disarrangement and/or dysregulation of intracellular signal trafficking subsequent to interaction with HCV characteristic proteins have been postulated to be a major pathogenic mechanism of the inflammatory changes (Bowen et al., 2005). One example of such research is Akt. It is a 57 kDa serine/threonine protein kinase B (PKB) expressed in fibroblasts, adipocytes and skeletal muscle, mediates many of the downstream events of phosphoinositide 3-kinase (PI3K) signaling. It is a critical intracellular signal activated by insulin and other growth factors and is also essential for most of the metabolic effects of insulin (Farese et al., 2005). Upon binding of insulin to its cell surface receptor, activation of its tyrosine kinase activity results in the phosphorylation of multiple substrates involved in its downstream effector functions. Activation of PI3K and phosphoinositide-dependent protein kinase signalling pathway results in the activation of multiple other related molecules, mediates intracellular signals to regulate a variety of cellular responses, including anti-apoptosis, proliferation, cell cycling, protein synthesis, glucose metabolism, and telomere activity (Noguchi et al., 2008). Experimental studies demonstrated that HCV proteins can activate PI3K/Akt pathway and expression of HCV core protein increased hepatic stellate cell (HSC) proliferation in a PI3K/Akt dependent fashion (Bataller et al., 2004). The increased phosphorylation correlated with increased IR to a variety of apoptotic stimuli (Banerjee et al., 2008; Street et al., 2005). PI3K/Akt signaling pathway was a concealed player involving in inflammatory process, thus contributed to HSC activation, liver fibrosis progression, and apotosis (Aleffi et al., 2005). Inhibition of PI3K signaling in HSC, which in turn reduced Akt activation, blocked the progression of liver fibrosis (Son et al., 2009). In addition, activation of the Akt signalling pathway has been demonstrated to contribute to hepatocarcinogenesis and predict outcomes of HCC in patients with HCV infection (Schmitz et al., 2008). Hepatic Akt expression of immunostaining was also significantly associated advanced fibrosis, a higher necroinflammatory activity, a lower BMI, a lower LDL-C and a lower gamma glutamyl transferase levels (Huang et al., 2011).

Since HCV-induced liver fibrosis has been demonstrated to be reversible after successful eradication of HCV infection, a long-term follow-up study with regard to the sequential change of Akt expression in patients receiving antiviral therapy is needed to elucidate this issue. In addition, activation of the Akt signalling pathway has been demonstrated to contribute to hepatocarcinogenesis in patients with HCV infection. Further follow-up study with respect to the correlation between Akt expression and liver cancer development also deserves.

4.4 Visfatin

Adipose tissue has emerged as a major endocrine organ. These adipokines play major roles in key aspects of metabolism, such as energy intake and expenditure, IR, fatty acid oxidation, inflammation, and immunity. Visfatin, as a visceral fat-specific adipokine, is a 52 kD protein that has been cloned already years ago as pre-B cell colony-enhancing factor (PBEF). Recently, visfatin was found to exert insulin-mimicking effects and may activate the intracellular signaling cascades for insulin(Fukuhara et al., 2005). Increasing visfatin level was significantly associated with T2DM(Chen et al., 2006; Dogru et al., 2007). Serum visfatin level was higher in obese than non-obese patients, and it was lower in non-alcoholic steatohepatitis patients than in those with simple steatosis or obese controls(Jarrar et al., 2008). In addition, visfatin level was shown to predict the presence of portal inflammation in non-alcoholic fatty liver disease patients(Aller et al., 2009). Visfatin was preferentially expressed by visceral adipose tissue compared with subcutaneous fat(Fukuhara et al., 2005). A positive correlation between visfatin and BMI as well as visceral fat mass has been reported(Hammarstedt et al., 2006). However, during the past years, contradictory results regarding waist circumference, glucose level, IR, visceral and subcutaneous adipose tissues have also been emerged in a clinical setting(Sommer et al., 2008). Berndt et al demonstrated that serum visfatin level did not correlate with visceral fat mass and that mRNA expression of the gene encoding visfatin was similar in visceral and subcutaneous adipose tissue (Berndt et al., 2005). Pagano et al demonstrated that serum visfatin level was decreased by approximately 50% in obese patients compared with controls. Furthermore, a negative correlation was found between serum visfatin level and BMI (Pagano et al., 2006). In the aspect of HCV infection, CHC patients tended to have a higher visfatin level than that of healthy controls. Serum visfatin level was correlated significantly with histological activity index scores and fibrosis stages, namely, disease severity in CHC patients (Huang et al., 2011). It hence implied that visfatin could be a potential biomarker for prediction of disease severity in HCV infection. The concordant results in different entities of liver diseases provided evidence for previous experimental studies indicating that visfatin was involved into inflammatory process network (Lim et al., 2008). On the other side, the studies addressing the correlation between visfatin and HCV genotypes were not concordant (Kukla et al., 2010). There was no significant correlation between visfatin and viral load, and treatment response to PegIFN/ribavirin therapy. Further study is warranted to clarify in which a concealed player or a bystander of the necroinflammatory and fibrogenetic scenarios visfatin behaves.

There was a tendency for higher visfatin levels in CHC patients with lower BMI (Huang et al., 2011; Kukla et al., 2010). MS was also negatively correlated with visfatin level in CHC patients. In addition, the extent of glucose abnormalities was not significantly correlated with visfatin level. These observations may imply that visfatin was not related to IR (Pagano et al., 2006). Kukla et al postulated that visfatin may play a dual role as a pro-inflammatory and/or protective factor (Kukla et al., 2010). Taken together, the speculations regarding the role of visfatin in CHC patients include 1) visfatin may not be one of the initiators in the context of metabolic derangements; 2) visfatin may not be involved into the link between obesity and IR in a clinical setting, at least among CHC patients. Therefore, its precise endocrine role deserved to be further clarified in HCV infection.

Several cytokines, particularly adipokines, have been investigated in terms of the link between HCV infection, steatosis, fibrosis, and metabolic abnormalities. These include leptin, adiponectin, apelin, resistin, TNF-α, interleukins, etc(Liu et al., 2005; Marra et al., 2009). It is informative and helpful for the elucidation of pathogenic mechanisms to assess the interaction between related biomarkers and clinical profiles of HCV infection, such as metabolic, virological, and histological factors.

5. Conclusion

Metabolic disorders are characteristic in patients with HCV infection. Besides established liver injury, the multiple extrahepatic metabolic disorders lead to a certain extent of impact on the disease activity, disease course, clinical outcomes, and treatment efficacy of current therapy. Enormous challenges for patient management have been arised in parallel with the promoter polymorphism at position -308 predicts response to combination therapy in hepatitis C virus infection. *J Infect Dis* 193(1), 98-101.

Dai, C. Y., Chuang, W. L., Ho, C. K., Hsieh, M. Y., Huang, J. F., Lee, L. P., Hou, N. J., Lin, Z. Y., Chen, S. C., Wang, L. Y., Tsai, J. F., Chang, W. Y., and Yu, M. L. (2008). Associations between hepatitis C viremia and low serum triglyceride and cholesterol levels: a community-based study. *J Hepatol* 49(1), 9-16.

Dai, C. Y., Huang, J. F., Hsieh, M. Y., Hou, N. J., Lin, Z. Y., Chen, S. C., Wang, L. Y., Chang, W. Y., Chuang, W. L., and Yu, M. L. (2009). Insulin resistance predicts response to peginterferon-alpha/ribavirin combination therapy in chronic hepatitis C patients. *J Hepatol* 50(4), 712-8.

Diercks, G. F., van Boven, A. J., Hillege, J. L., de Jong, P. E., Rouleau, J. L., and van Gilst, W. H. (2002). The importance of microalbuminuria as a cardiovascular risk indicator: A review. *Can J Cardiol* 18(5), 525-35.

Dogru, T., Sonmez, A., Tasci, I., Bozoglu, E., Yilmaz, M. I., Genc, H., Erdem, G., Gok, M., Bingol, N., Kilic, S., Ozgurtas, T., and Bingol, S. (2007). Plasma visfatin levels in patients with newly diagnosed and untreated type 2 diabetes mellitus and impaired glucose tolerance. *Diabetes Res Clin Pract* 76(1), 24-9.

Erikstrup, C., Mortensen, O. H., and Pedersen, B. K. (2006). Retinol-binding protein 4 and insulin resistance. *N Engl J Med* 355(13), 1393-4; author reply 1394-5.

Farese, R. V., Sajan, M. P., and Standaert, M. L. (2005). Atypical protein kinase C in insulin action and insulin resistance. *Biochem Soc Trans* 33(Pt 2), 350-3.

Floris-Moore, M., Howard, A. A., Lo, Y., Schoenbaum, E. E., Arnsten, J. H., and Klein, R. S. (2007). Hepatitis C infection is associated with lower lipids and high-sensitivity C-reactive protein in HIV-infected men. *AIDS Patient Care STDS* 21(7), 479-91.

Fried, M. W., Shiffman, M. L., Reddy, K. R., Smith, C., Marinos, G., Goncales, F. L., Jr., Haussinger, D., Diago, M., Carosi, G., Dhumeaux, D., Craxi, A., Lin, A., Hoffman, J., and Yu, J. (2002). Peginterferon alfa-2a plus ribavirin for chronic hepatitis C virus infection. *N Engl J Med* 347(13), 975-82.

Fukuhara, A., Matsuda, M., Nishizawa, M., Segawa, K., Tanaka, M., Kishimoto, K., Matsuki, Y., Murakami, M., Ichisaka, T., Murakami, H., Watanabe, E., Takagi, T., Akiyoshi, M., Ohtsubo, T., Kihara, S., Yamashita, S., Makishima, M., Funahashi, T., Yamanaka, S., Hiramatsu, R., Matsuzawa, Y., and Shimomura, I. (2005). Visfatin: a protein secreted by visceral fat that mimics the effects of insulin. *Science* 307(5708), 426-30.

Garini, G., Allegri, L., Vaglio, A., and Buzio, C. (2005). Hepatitis C virus-related cryoglobulinemia and glomerulonephritis: pathogenesis and therapeutic strategies. *Ann Ital Med Int* 20(2), 71-80.

Gerstein, H. C., Yusuf, S., Bosch, J., Pogue, J., Sheridan, P., Dinccag, N., Hanefeld, M., Hoogwerf, B., Laakso, M., Mohan, V., Shaw, J., Zinman, B., and Holman, R. R. (2006). Effect of rosiglitazone on the frequency of diabetes in patients with impaired glucose tolerance or impaired fasting glucose: a randomised controlled trial. *Lancet* 368(9541), 1096-105.

Graham, T. E., Yang, Q., Bluher, M., Hammarstedt, A., Ciaraldi, T. P., Henry, R. R., Wason, C. J., Oberbach, A., Jansson, P. A., Smith, U., and Kahn, B. B. (2006). Retinol-binding protein 4 and insulin resistance in lean, obese, and diabetic subjects. *N Engl J Med* 354(24), 2552-63.

Grundy, S. M., Cleeman, J. I., Daniels, S. R., Donato, K. A., Eckel, R. H., Franklin, B. A., Gordon, D. J., Krauss, R. M., Savage, P. J., Smith, S. C., Jr., Spertus, J. A., and Costa, F. (2005). Diagnosis and management of the metabolic syndrome: an American Heart Association/National Heart, Lung, and Blood Institute Scientific Statement. *Circulation* 112(17), 2735-52.

Hadziyannis, S. J. (1997). The spectrum of extrahepatic manifestations in hepatitis C virus infection. *J Viral Hepat* 4(1), 9-28.

Hammarstedt, A., Pihlajamaki, J., Rotter Sopasakis, V., Gogg, S., Jansson, P. A., Laakso, M., and Smith, U. (2006). Visfatin is an adipokine, but it is not regulated by thiazolidinediones. *J Clin Endocrinol Metab* 91(3), 1181-4.

Harris, M. I., Klein, R., Welborn, T. A., and Knuiman, M. W. (1992). Onset of NIDDM occurs at least 4-7 yr before clinical diagnosis. *Diabetes Care* 15(7), 815-9.

Heish, M. Y., Dai, C. Y., Huang, J. F., Lee, L. P., Hou, N. J., Chuang, W. L., and Yu, M. L. (2008). Revisit of oral glucose tolerance test: a must for diagnosis of type 2 diabetes in patients with chronic hepatitis C. *Am J Gastroenterol* 103(2), 487-8; author reply 488.

Hsieh, M. H., Lee, L. P., Hsieh, M. Y., Tsai, K. B., Huang, J. F., Hou, N. J., Chen, S. C., Lin, Z. Y., Wang, L. Y., Dai, C. Y., Chuang, W. L., and Yu, M. L. (2007). Hepatic steatosis and fibrosis in chronic hepatitis C in Taiwan. *Jpn J Infect Dis* 60(6), 377-81.

Hsieh, M. Y., Lee, L. P., Hou, N. J., Yang, J. F., Huang, J. F., Dai, C. Y., Chuang, W. L., Lin, Z. Y., Chen, S. C., Wang, L. Y., Chang, W. Y., and Yu, M. L. (2007). Qualitative application of COBAS AMPLICOR HCV test version 2.0 assays in patients with chronic hepatitis C virus infection and comparison of clinical performance with version 1.0. *Kaohsiung J Med Sci* 23(7), 332-8.

Hsu, C. S., Liu, C. J., Liu, C. H., Wang, C. C., Chen, C. L., Lai, M. Y., Chen, P. J., Kao, J. H., and Chen, D. S. (2008). High hepatitis C viral load is associated with insulin resistance in patients with chronic hepatitis C. *Liver Int* 28(2), 271-7.

Huang, J. F., Chuang, W. L., Dai, C. Y., Ho, C. K., Hwang, S. J., Chen, S. C., Lin, Z. Y., Wang, L. Y., Chang, W. Y., and Yu, M. L. (2006). Viral hepatitis and proteinuria in an area endemic for hepatitis B and C infections: another chain of link? *J Intern Med* 260(3), 255-62.

Huang, J. F., Chuang, W. L., Yu, M. L., Yu, S. H., Huang, C. F., Huang, C. I., Yeh, M. L., Hsieh, M. H., Yang, J. F., Lin, Z. Y., Chen, S. C., Dai, C. Y., and Chang, W. Y. (2009). Hepatitis C virus infection and metabolic syndrome---a community-based study in an endemic area of Taiwan. *Kaohsiung J Med Sci* 25(6), 299-305.

Huang, J. F., Chuang, Y. H., Dai, C. Y., Yu, M. L., Huang, C. F., Hsiao, P. J., Hsieh, M. Y., Huang, C. I., Yeh, M. L., Yang, J. F., Lin, Z. Y., Chen, S. C., and Chuang, W. L. (2011). Hepatic Akt expression correlates with advanced fibrosis in patients with chronic hepatitis C infection. *Hepatol Res* 41(5), 430-436.

Huang, J. F., Dai, C. Y., Hwang, S. J., Ho, C. K., Hsiao, P. J., Hsieh, M. Y., Lee, L. P., Lin, Z. Y., Chen, S. C., Wang, L. Y., Shin, S. J., Chang, W. Y., Chuang, W. L., and Yu, M. L. (2007). Hepatitis C viremia increases the association with type 2 diabetes mellitus in a hepatitis B and C endemic area: an epidemiological link with virological implication. *Am J Gastroenterol* 102(6), 1237-43.

Huang, J. F., Dai, C. Y., Yu, M. L., Huang, C. F., Huang, C. I., Yeh, M. L., Yang, J. F., Hou, N. J., Hsiao, P. J., Lin, Z. Y., Chen, S. C., Shin, S. J., and Chuang, W. L. (2011b).

Pegylated interferon plus ribavirin therapy improves pancreatic beta-cell function in chronic hepatitis C patients. *Liver Int.*

Huang, J. F., Dai, C. Y., Yu, M. L., Shin, S. J., Hsieh, M. Y., Huang, C. F., Lee, L. P., Lin, K. D., Lin, Z. Y., Chen, S. C., Hsieh, M. Y., Wang, L. Y., Chang, W. Y., and Chuang, W. L. (2009). Serum retinol-binding protein 4 is inversely correlated with disease severity of chronic hepatitis C. *J Hepatol* 50(3), 471-8.

Huang, J. F., Huang, C. F., Yu, M. L., Dai, C. Y., Huang, C. I., Yeh, M. L., Hsieh, M. H., Yang, J. F., Hsieh, M. Y., Lin, Z. Y., Chen, S. C., and Chuang, W. L. (2011). Serum visfatin is correlated with disease severity and metabolic syndrome in chronic hepatitis C infection. *J Gastroenterol Hepatol* 26(3), 530-5.

Huang, J. F., Yu, M. L., Dai, C. Y., Hsieh, M. Y., Hwang, S. J., Hsiao, P. J., Lee, L. P., Lin, Z. Y., Chen, S. C., Wang, L. Y., Shin, S. J., Chang, W. Y., and Chuang, W. L. (2008). Reappraisal of the characteristics of glucose abnormalities in patients with chronic hepatitis C infection. *Am J Gastroenterol* 103(8), 1933-40.

Hui, J. M., Sud, A., Farrell, G. C., Bandara, P., Byth, K., Kench, J. G., McCaughan, G. W., and George, J. (2003). Insulin resistance is associated with chronic hepatitis C virus infection and fibrosis progression [corrected]. *Gastroenterology* 125(6), 1695-704.

Ishizaka, N., Ishizaka, Y., Takahashi, E., Tooda, E., Hashimoto, H., Nagai, R., and Yamakado, M. (2002). Association between hepatitis C virus seropositivity, carotid-artery plaque, and intima-media thickening. *Lancet* 359(9301), 133-5.

Ito, H., Hattori, S., Matusda, I., Amamiya, S., Hajikano, H., Yoshizawa, H., Miyakawa, Y., and Mayumi, M. (1981). Hepatitis B e antigen-mediated membranous glomerulonephritis. Correlation of ultrastructural changes with HBeAg in the serum and glomeruli. *Lab Invest* 44(3), 214-20.

Iwasa, M., Hara, N., Miyachi, H., Tanaka, H., Takeo, M., Fujita, N., Kobayashi, Y., Kojima, Y., Kaito, M., and Takei, Y. (2009). Patients achieving clearance of HCV with interferon therapy recover from decreased retinol-binding protein 4 levels. *J Viral Hepat.*

Janke, J., Engeli, S., Boschmann, M., Adams, F., Bohnke, J., Luft, F. C., Sharma, A. M., and Jordan, J. (2006). Retinol-binding protein 4 in human obesity. *Diabetes* 55(10), 2805-10.

Jarrar, M. H., Baranova, A., Collantes, R., Ranard, B., Stepanova, M., Bennett, C., Fang, Y., Elariny, H., Goodman, Z., Chandhoke, V., and Younossi, Z. M. (2008). Adipokines and cytokines in non-alcoholic fatty liver disease. *Aliment Pharmacol Ther* 27(5), 412-21.

Johnson, R. J., and Couser, W. G. (1990). Hepatitis B infection and renal disease: clinical, immunopathogenetic and therapeutic considerations. *Kidney Int* 37(2), 663-76.

Kalabay, L., Nemesanszky, E., Csepregi, A., Pusztay, M., David, K., Horvath, G., Ibranyi, E., Telegdy, L., Par, A., Biro, A., Fekete, B., Gervain, J., Horanyi, M., Ribiczey, P., Csondes, M., Kleiber, M., Walentin, S., Prohaszka, Z., and Fust, G. (2004). Paradoxical alteration of acute-phase protein levels in patients with chronic hepatitis C treated with IFN-alpha2b. *Int Immunol* 16(1), 51-4.

Kannel, W. B., and McGee, D. L. (1979). Diabetes and glucose tolerance as risk factors for cardiovascular disease: the Framingham study. *Diabetes Care* 2(2), 120-6.

Kawaguchi, T., Ide, T., Taniguchi, E., Hirano, E., Itou, M., Sumie, S., Nagao, Y., Yanagimoto, C., Hanada, S., Koga, H., and Sata, M. (2007). Clearance of HCV improves insulin resistance, beta-cell function, and hepatic expression of insulin receptor substrate 1 and 2. *Am J Gastroenterol* 102(3), 570-6.

Kawaguchi, T., Yoshida, T., Harada, M., Hisamoto, T., Nagao, Y., Ide, T., Taniguchi, E., Kumemura, H., Hanada, S., Maeyama, M., Baba, S., Koga, H., Kumashiro, R., Ueno, T., Ogata, H., Yoshimura, A., and Sata, M. (2004). Hepatitis C virus down-regulates insulin receptor substrates 1 and 2 through up-regulation of suppressor of cytokine signaling 3. *Am J Pathol* 165(5), 1499-508.

Keane, J., Meier, J. L., Noth, R. H., and Swislocki, A. L. (2009). Computer-based screening of veterans for metabolic syndrome. *Metab Syndr Relat Disord* 7(6), 557-61.

Keane, W. F. (2000). Proteinuria: its clinical importance and role in progressive renal disease. *Am J Kidney Dis* 35(4 Suppl 1), S97-105.

Keeffe, E. B. (2005). Chronic hepatitis C: management of treatment failures. *Clin Gastroenterol Hepatol* 3(10 Suppl 2), S102-5.

Knobler, H., Schihmanter, R., Zifroni, A., Fenakel, G., and Schattner, A. (2000). Increased risk of type 2 diabetes in noncirrhotic patients with chronic hepatitis C virus infection. *Mayo Clin Proc* 75(4), 355-9.

Knowler, W. C., Barrett-Connor, E., Fowler, S. E., Hamman, R. F., Lachin, J. M., Walker, E. A., and Nathan, D. M. (2002). Reduction in the incidence of type 2 diabetes with lifestyle intervention or metformin. *N Engl J Med* 346(6), 393-403.

Koenig, W., Khuseyinova, N., Baumert, J., and Meisinger, C. (2008). Prospective study of high-sensitivity C-reactive protein as a determinant of mortality: results from the MONICA/KORA Augsburg Cohort Study, 1984-1998. *Clin Chem* 54(2), 335-42.

Kogiso, T., Moriyoshi, Y., Shimizu, S., Nagahara, H., and Shiratori, K. (2009). High-sensitivity C-reactive protein as a serum predictor of nonalcoholic fatty liver disease based on the Akaike Information Criterion scoring system in the general Japanese population. *J Gastroenterol* 44(4), 313-21.

Kukla, M., Zwirska-Korczala, K., Gabriel, A., Waluga, M., Warakomska, I., Berdowska, A., Rybus-Kalinowska, B., Kalinowski, M., Janczewska-Kazek, E., Wozniak-Grygiel, E., and Kryczka, W. (2010). Visfatin serum levels in chronic hepatitis C patients. *J Viral Hepat* 17(4), 254-60.

Kushner, I. (1982). The phenomenon of the acute phase response. *Ann N Y Acad Sci* 389, 39-48.

Lecube, A., Hernandez, C., Genesca, J., Esteban, J. I., Jardi, R., and Simo, R. (2004). High prevalence of glucose abnormalities in patients with hepatitis C virus infection: a multivariate analysis considering the liver injury. *Diabetes Care* 27(5), 1171-5.

Lecube, A., Hernandez, C., Genesca, J., and Simo, R. (2006). Glucose abnormalities in patients with hepatitis C virus infection: Epidemiology and pathogenesis. *Diabetes Care* 29(5), 1140-9.

Lecube, A., Hernandez, C., Genesca, J., and Simo, R. (2006). Proinflammatory cytokines, insulin resistance, and insulin secretion in chronic hepatitis C patients: A case-control study. *Diabetes Care* 29(5), 1096-101.

Lee, J. J., Lin, M. Y., Yang, Y. H., Lu, S. N., Chen, H. C., and Hwang, S. J. (2010). Association of hepatitis C and B virus infection with CKD in an endemic area in Taiwan: a cross-sectional study. *Am J Kidney Dis* 56(1), 23-31.

Levo, Y., Gorevic, P. D., Kassab, H. J., Zucker-Franklin, D., and Franklin, E. C. (1977). Association between hepatitis B virus and essential mixed cryoglobulinemia. *N Engl J Med* 296(26), 1501-4.

Liangpunsakul, S., and Chalasani, N. (2005). Relationship between hepatitis C and microalbuminuria: results from the NHANES III. *Kidney Int* 67(1), 285-90.

Lim, S. Y., Davidson, S. M., Paramanathan, A. J., Smith, C. C., Yellon, D. M., and Hausenloy, D. J. (2008). The novel adipocytokine visfatin exerts direct cardioprotective effects. *J Cell Mol Med* 12(4), 1395-403.

Liu, C. J., Chen, P. J., Jeng, Y. M., Huang, W. L., Yang, W. S., Lai, M. Y., Kao, J. H., and Chen, D. S. (2005). Serum adiponectin correlates with viral characteristics but not histologic features in patients with chronic hepatitis C. *J Hepatol* 43(2), 235-42.

Marra, F., and Bertolani, C. (2009). Adipokines in liver diseases. *Hepatology* 50(3), 957-69.

Mason, A. L., Lau, J. Y., Hoang, N., Qian, K., Alexander, G. J., Xu, L., Guo, L., Jacob, S., Regenstein, F. G., Zimmerman, R., Everhart, J. E., Wasserfall, C., Maclaren, N. K., and Perrillo, R. P. (1999). Association of diabetes mellitus and chronic hepatitis C virus infection. *Hepatology* 29(2), 328-33.

Matthews, D. R., Hosker, J. P., Rudenski, A. S., Naylor, B. A., Treacher, D. F., and Turner, R. C. (1985). Homeostasis model assessment: insulin resistance and beta-cell function from fasting plasma glucose and insulin concentrations in man. *Diabetologia* 28(7), 412-9.

Mazzaro, C., Panarello, G., Tesio, F., Santini, G., Crovatto, M., Mazzi, G., Zorat, F., Tulissi, P., Pussini, E., Baracetti, S., Campanacci, L., and Pozzato, G. (2000). Hepatitis C virus risk: a hepatitis C virus related syndrome. *J Intern Med* 247(5), 535-45.

Mehta, S. H., Brancati, F. L., Strathdee, S. A., Pankow, J. S., Netski, D., Coresh, J., Szklo, M., and Thomas, D. L. (2003). Hepatitis C virus infection and incident type 2 diabetes. *Hepatology* 38(1), 50-6.

Mehta, S. H., Brancati, F. L., Sulkowski, M. S., Strathdee, S. A., Szklo, M., and Thomas, D. L. (2000). Prevalence of type 2 diabetes mellitus among persons with hepatitis C virus infection in the United States. *Ann Intern Med* 133(8), 592-9.

Meyers, C. M., Seeff, L. B., Stehman-Breen, C. O., and Hoofnagle, J. H. (2003). Hepatitis C and renal disease: an update. *Am J Kidney Dis* 42(4), 631-57.

Monazahian, M., Bohme, I., Bonk, S., Koch, A., Scholz, C., Grethe, S., and Thomssen, R. (1999). Low density lipoprotein receptor as a candidate receptor for hepatitis C virus. *J Med Virol* 57(3), 223-9.

Monazahian, M., Kippenberger, S., Muller, A., Seitz, H., Bohme, I., Grethe, S., and Thomssen, R. (2000). Binding of human lipoproteins (low, very low, high density lipoproteins) to recombinant envelope proteins of hepatitis C virus. *Med Microbiol Immunol* 188(4), 177-84.

Moritani, M., Adachi, K., Arima, N., Takashima, T., Miyaoka, Y., Niigaki, M., Furuta, K., Sato, S., and Kinoshita, Y. (2005). A study of arteriosclerosis in healthy subjects with HBV and HCV infection. *J Gastroenterol* 40(11), 1049-53.

Moriya, K., Shintani, Y., Fujie, H., Miyoshi, H., Tsutsumi, T., Yotsuyanagi, H., Iino, S., Kimura, S., and Koike, K. (2003). Serum lipid profile of patients with genotype 1b hepatitis C viral infection in Japan. *Hepatol Res* 25(4), 371-376.

Moucari, R., Marcellin, P., and Asselah, T. (2007). [Steatosis during chronic hepatitis C: the role of insulin resistance and viral factors]. *Gastroenterol Clin Biol* 31(8-9 Pt 1), 643-54.

Muir, A. J., Bornstein, J. D., and Killenberg, P. G. (2004). Peginterferon alfa-2b and ribavirin for the treatment of chronic hepatitis C in blacks and non-Hispanic whites. *N Engl J Med* 350(22), 2265-71.

Muramatsu, T., Hora, K., Ako, S., Tachibana, N., and Tanaka, E. (2000). The role of hepatitis C virus infection in glomerulopathy. *Hepatol Res* 18(3), 190-202.

Muzzi, A., Leandro, G., Rubbia-Brandt, L., James, R., Keiser, O., Malinverni, R., Dufour, J. F., Helbling, B., Hadengue, A., Gonvers, J. J., Mullhaupt, B., Cerny, A., Mondelli, M. U., and Negro, F. (2005). Insulin resistance is associated with liver fibrosis in non-diabetic chronic hepatitis C patients. *J Hepatol* 42(1), 41-6.

Nascimento, M. M., Bruchfeld, A., Suliman, M. E., Hayashi, S. Y., Pecoits-Filho, R., Manfro, R. C., Pachaly, M. A., Renner, L., Stenvinkel, P., Riella, M. C., and Lindholm, B. (2005). Effect of hepatitis C serology on C-reactive protein in a cohort of Brazilian hemodialysis patients. *Braz J Med Biol Res* 38(5), 783-8.

Negro, F. (2010). Abnormalities of lipid metabolism in hepatitis C virus infection. *Gut* 59(9), 1279-87.

Neuschwander-Tetri, B. A. (2008). Hepatitis C virus-induced insulin resistance: not all genotypes are the same. *Gastroenterology* 134(2), 619-22.

Newcomer, M. E., and Ong, D. E. (2000). Plasma retinol binding protein: structure and function of the prototypic lipocalin. *Biochim Biophys Acta* 1482(1-2), 57-64.

Noguchi, M., and Kinowaki, K. (2008). [PI3K-AKT network roles in infectious diseases]. *Kansenshogaku Zasshi* 82(3), 161-7.

Oyake, N., Shimada, T., Murakami, Y., Ishibashi, Y., Satoh, H., Suzuki, K., Matsumory, A., and Oda, T. (2008). Hepatitis C virus infection as a risk factor for increased aortic stiffness and cardiovascular events in dialysis patients. *J Nephrol* 21(3), 345-53.

Pagano, C., Pilon, C., Olivieri, M., Mason, P., Fabris, R., Serra, R., Milan, G., Rossato, M., Federspil, G., and Vettor, R. (2006). Reduced plasma visfatin/pre-B cell colony-enhancing factor in obesity is not related to insulin resistance in humans. *J Clin Endocrinol Metab* 91(8), 3165-70.

Perlemuter, G., Sabile, A., Letteron, P., Vona, G., Topilco, A., Chretien, Y., Koike, K., Pessayre, D., Chapman, J., Barba, G., and Brechot, C. (2002). Hepatitis C virus core protein inhibits microsomal triglyceride transfer protein activity and very low density lipoprotein secretion: a model of viral-related steatosis. *FASEB J* 16(2), 185-94.

Petit, J. M., Bour, J. B., Galland-Jos, C., Minello, A., Verges, B., Guiguet, M., Brun, J. M., and Hillon, P. (2001). Risk factors for diabetes mellitus and early insulin resistance in chronic hepatitis C. *J Hepatol* 35(2), 279-83.

Petta, S., Camma, C., Di Marco, V., Alessi, N., Barbaria, F., Cabibi, D., Caldarella, R., Ciminnisi, S., Licata, A., Massenti, M. F., Mazzola, A., Tarantino, G., Marchesini, G., and Craxi, A. (2008). Retinol-binding protein 4: a new marker of virus-induced steatosis in patients infected with hepatitis c virus genotype 1. *Hepatology* 48(1), 28-37.

Poustchi, H., Negro, F., Hui, J., Cua, I. H., Brandt, L. R., Kench, J. G., and George, J. (2008). Insulin resistance and response to therapy in patients infected with chronic hepatitis C virus genotypes 2 and 3. *J Hepatol* 48(1), 28-34.

Powell, E. E., Jonsson, J. R., and Clouston, A. D. (2005). Steatosis: co-factor in other liver diseases. *Hepatology* 42(1), 5-13.

Ramadori, G., and Armbrust, T. (2001). Cytokines in the liver. *Eur J Gastroenterol Hepatol* 13(7), 777-84.

Reingold, J., Wanke, C., Kotler, D., Lewis, C., Tracy, R., Heymsfield, S., Tien, P., Bacchetti, P., Scherzer, R., Grunfeld, C., and Shlipak, M. (2008). Association of HIV infection and HIV/HCV coinfection with C-reactive protein levels: the fat redistribution and metabolic change in HIV infection (FRAM) study. *J Acquir Immune Defic Syndr* 48(2), 142-8.

Ridker, P. M., Buring, J. E., Cook, N. R., and Rifai, N. (2003). C-reactive protein, the metabolic syndrome, and risk of incident cardiovascular events: an 8-year follow-up of 14 719 initially healthy American women. *Circulation* 107(3), 391-7.

Riquelme, A., Arrese, M., Soza, A., Morales, A., Baudrand, R., Perez-Ayuso, R. M., Gonzalez, R., Alvarez, M., Hernandez, V., Garcia-Zattera, M. J., Otarola, F., Medina, B., Rigotti, A., Miquel, J. F., Marshall, G., and Nervi, F. (2009). Non-alcoholic fatty liver disease and its association with obesity, insulin resistance and increased serum levels of C-reactive protein in Hispanics. *Liver Int* 29(1), 82-8.

Romero-Gomez, M., Del Mar Viloria, M., Andrade, R. J., Salmeron, J., Diago, M., Fernandez-Rodriguez, C. M., Corpas, R., Cruz, M., Grande, L., Vazquez, L., Munoz-De-Rueda, P., Lopez-Serrano, P., Gila, A., Gutierrez, M. L., Perez, C., Ruiz-Extremera, A., Suarez, E., and Castillo, J. (2005). Insulin resistance impairs sustained response rate to peginterferon plus ribavirin in chronic hepatitis C patients. *Gastroenterology* 128(3), 636-41.

Saltiel, A. R., and Kahn, C. R. (2001). Insulin signalling and the regulation of glucose and lipid metabolism. *Nature* 414(6865), 799-806.

Sanyal, A. J. (2005). Review article: non-alcoholic fatty liver disease and hepatitis C--risk factors and clinical implications. *Aliment Pharmacol Ther* 22 Suppl 2, 48-51.

Schmitz, K. J., Wohlschlaeger, J., Lang, H., Sotiropoulos, G. C., Malago, M., Steveling, K., Reis, H., Cicinnati, V. R., Schmid, K. W., and Baba, H. A. (2008). Activation of the ERK and AKT signalling pathway predicts poor prognosis in hepatocellular carcinoma and ERK activation in cancer tissue is associated with hepatitis C virus infection. *J Hepatol* 48(1), 83-90.

Serfaty, L., Andreani, T., Giral, P., Carbonell, N., Chazouilleres, O., and Poupon, R. (2001). Hepatitis C virus induced hypobetalipoproteinemia: a possible mechanism for steatosis in chronic hepatitis C. *J Hepatol* 34(3), 428-34.

Serste, T., Nkuize, M., Moucari, R., Van Gossum, M., Reynders, M., Scheen, R., Vertongen, F., Buset, M., Mulkay, J. P., and Marcellin, P. (2010). Metabolic disorders associated with chronic hepatitis C: impact of genotype and ethnicity. *Liver Int* 30(8), 1131-6.

Shintani, Y., Fujie, H., Miyoshi, H., Tsutsumi, T., Tsukamoto, K., Kimura, S., Moriya, K., and Koike, K. (2004). Hepatitis C virus infection and diabetes: direct involvement of the virus in the development of insulin resistance. *Gastroenterology* 126(3), 840-8.

Siagris, D., Christofidou, M., Theocharis, G. J., Pagoni, N., Papadimitriou, C., Lekkou, A., Thomopoulos, K., Starakis, I., Tsamandas, A. C., and Labropoulou-Karatza, C. (2006). Serum lipid pattern in chronic hepatitis C: histological and virological correlations. *J Viral Hepat* 13(1), 56-61.

Sommer, G., Garten, A., Petzold, S., Beck-Sickinger, A. G., Bluher, M., Stumvoll, M., and Fasshauer, M. (2008). Visfatin/PBEF/Nampt: structure, regulation and potential function of a novel adipokine. *Clin Sci (Lond)* 115(1), 13-23.

Son, G., Hines, I. N., Lindquist, J., Schrum, L. W., and Rippe, R. A. (2009). Inhibition of phosphatidylinositol 3-kinase signaling in hepatic stellate cells blocks the progression of hepatic fibrosis. *Hepatology* 50(5), 1512-23.

Strader, D. B., Wright, T., Thomas, D. L., and Seeff, L. B. (2004). Diagnosis, management, and treatment of hepatitis C. *Hepatology* 39(4), 1147-71.

Strassburg, C. P., Obermayer-Straub, P., and Manns, M. P. (1996). Autoimmunity in hepatitis C and D virus infection. *J Viral Hepat* 3(2), 49-59.

Street, A., Macdonald, A., McCormick, C., and Harris, M. (2005). Hepatitis C virus NS5A-mediated activation of phosphoinositide 3-kinase results in stabilization of cellular beta-catenin and stimulation of beta-catenin-responsive transcription. *J Virol* 79(8), 5006-16.

Sud, A., Hui, J. M., Farrell, G. C., Bandara, P., Kench, J. G., Fung, C., Lin, R., Samarasinghe, D., Liddle, C., McCaughan, G. W., and George, J. (2004). Improved prediction of fibrosis in chronic hepatitis C using measures of insulin resistance in a probability index. *Hepatology* 39(5), 1239-47.

Svegliati-Baroni, G., Bugianesi, E., Bouserhal, T., Marini, F., Ridolfi, F., Tarsetti, F., Ancarani, F., Petrelli, E., Peruzzi, E., Lo Cascio, M., Rizzetto, M., Marchesini, G., and Benedetti, A. (2007). Post-load insulin resistance is an independent predictor of hepatic fibrosis in virus C chronic hepatitis and in non-alcoholic fatty liver disease. *Gut* 56(9), 1296-301.

Tada, S., Saito, H., Ebinuma, H., Ojiro, K., Yamagishi, Y., Kumagai, N., Inagaki, Y., Masuda, T., Nishida, J., Takahashi, M., Nagata, H., and Hibi, T. (2009). Treatment of hepatitis C virus with peg-interferon and ribavirin combination therapy significantly affects lipid metabolism. *Hepatol Res* 39(2), 195-9.

Tai, T. Y., Lu, J. Y., Chen, C. L., Lai, M. Y., Chen, P. J., Kao, J. H., Lee, C. Z., Lee, H. S., Chuang, L. M., and Jeng, Y. M. (2003). Interferon-alpha reduces insulin resistance and beta-cell secretion in responders among patients with chronic hepatitis B and C. *J Endocrinol* 178(3), 457-65.

Takashima, N., Tomoike, H., and Iwai, N. (2006). Retinol-binding protein 4 and insulin resistance. *N Engl J Med* 355(13), 1392; author reply 1394-5.

Takekoshi, Y., Tanaka, M., Miyakawa, Y., Yoshizawa, H., Takahashi, K., and Mayumi, M. (1979). Free "small" and IgG-associated "large" hepatitis B e antigen in the serum and glomerular capillary walls of two patients with membranous glomerulonephritis. *N Engl J Med* 300(15), 814-9.

Tamori, Y., Sakaue, H., and Kasuga, M. (2006). RBP4, an unexpected adipokine. *Nat Med* 12(1), 30-1; discussion 31.

Tarantino, G., Conca, P., Sorrentino, P., and Ariello, M. (2006). Metabolic factors involved in the therapeutic response of patients with hepatitis C virus-related chronic hepatitis. *J Gastroenterol Hepatol* 21(8), 1266-8.

Taura, N., Ichikawa, T., Hamasaki, K., Nakao, K., Nishimura, D., Goto, T., Fukuta, M., Kawashimo, H., Fujimoto, M., Kusumoto, K., Motoyoshi, Y., Shibata, H., Abiru, N., Yamasaki, H., and Eguchi, K. (2006). Association between liver fibrosis and insulin sensitivity in chronic hepatitis C patients. *Am J Gastroenterol* 101(12), 2752-9.

Thompson, A. J., Patel, K., Chuang, W. L., Lawitz, E. J., Rodriguez-Torres, M., Rustgi, V. K., Flisiak, R., Pianko, S., Diago, M., Arora, S., Foster, G. R., Torbenson, M., Benhamou, Y., Nelson, D. R., Sulkowski, M. S., Zeuzem, S., Pulkstenis, E., Subramanian, G. M., and McHutchison, J. G. (2012). Viral clearance is associated with improved insulin resistance in genotype 1 chronic hepatitis C but not genotype 2/3. Gut 61(1):128-34.

Tsui, J. I., Whooley, M. A., Monto, A., Seal, K., Tien, P. C., and Shlipak, M. (2009). Association of hepatitis C virus seropositivity with inflammatory markers and heart failure in persons with coronary heart disease: data from the Heart and Soul study. *J Card Fail* 15(5), 451-6.

Tuomilehto, J., Lindstrom, J., Eriksson, J. G., Valle, T. T., Hamalainen, H., Ilanne-Parikka, P., Keinanen-Kiukaanniemi, S., Laakso, M., Louheranta, A., Rastas, M., Salminen, V., and Uusitupa, M. (2001). Prevention of type 2 diabetes mellitus by changes in lifestyle among subjects with impaired glucose tolerance. *N Engl J Med* 344(18), 1343-50.

von Eynatten, M., Lepper, P. M., Liu, D., Lang, K., Baumann, M., Nawroth, P. P., Bierhaus, A., Dugi, K. A., Heemann, U., Allolio, B., and Humpert, P. M. (2007). Retinol-binding protein 4 is associated with components of the metabolic syndrome, but not with insulin resistance, in men with type 2 diabetes or coronary artery disease. *Diabetologia* 50(9), 1930-7.

Wang, C. S., Wang, S. T., Yao, W. J., Chang, T. T., and Chou, P. (2003). Community-based study of hepatitis C virus infection and type 2 diabetes: an association affected by age and hepatitis severity status. *Am J Epidemiol* 158(12), 1154-60.

Wang, C. S., Wang, S. T., Yao, W. J., Chang, T. T., and Chou, P. (2007). Hepatitis C virus infection and the development of type 2 diabetes in a community-based longitudinal study. *Am J Epidemiol* 166(2), 196-203.

Watanabe, S., Yaginuma, R., Ikejima, K., and Miyazaki, A. (2008). Liver diseases and metabolic syndrome. *J Gastroenterol* 43(7), 509-18.

Yagmur, E., Weiskirchen, R., Gressner, A. M., Trautwein, C., and Tacke, F. (2007). Insulin resistance in liver cirrhosis is not associated with circulating retinol-binding protein 4. *Diabetes Care* 30(5), 1168-72.

Yang, J. F., Dai, C. Y., Chuang, W. L., Lin, W. Y., Lin, Z. Y., Chen, S. C., Hsieh, M. Y., Wang, L. Y., Tsai, J. F., Chang, W. Y., and Yu, M. L. (2006). Prevalence and clinical significance of HGV/GBV-C infection in patients with chronic hepatitis B or C. *Jpn J Infect Dis* 59(1), 25-30.

Yang, Q., Graham, T. E., Mody, N., Preitner, F., Peroni, O. D., Zabolotny, J. M., Kotani, K., Quadro, L., and Kahn, B. B. (2005). Serum retinol binding protein 4 contributes to insulin resistance in obesity and type 2 diabetes. *Nature* 436(7049), 356-62.

Yelken, B., Gorgulu, N., Caliskan, Y., Elitok, A., Cimen, A. O., Yazici, H., Oflaz, H., Turkmen, A., and Sever, M. S. (2009). Association between chronic hepatitis C infection and coronary flow reserve in dialysis patients with failed renal allografts. *Transplant Proc* 41(5), 1519-23.

Yu, M. L., and Chuang, W. L. (2009). Treatment of chronic hepatitis C in Asia: when East meets West. *J Gastroenterol Hepatol* 24(3), 336-45.

Yu, M. L., Chuang, W. L., Chen, S. C., Dai, C. Y., Hou, C., Wang, J. H., Lu, S. N., Huang, J. F., Lin, Z. Y., Hsieh, M. Y., Tsai, J. F., Wang, L. Y., and Chang, W. Y. (2001a). Changing prevalence of hepatitis C virus genotypes: molecular epidemiology and clinical implications in the hepatitis C virus hyperendemic areas and a tertiary referral center in Taiwan. *J Med Virol* 65(1), 58-65.

Yu, M. L., Chuang, W. L., Dai, C. Y., Chen, S. C., Lin, Z. Y., Hsieh, M. Y., Tsai, J. F., Wang, L. Y., and Chang, W. Y. (2001). GB virus C/hepatitis G virus infection in chronic hepatitis C patients with and without interferon-alpha therapy. *Antiviral Res* 52(3), 241-9.

Yu, M. L., Dai, C. Y., Huang, J. F., Chiu, C. F., Yang, Y. H., Hou, N. J., Lee, L. P., Hsieh, M. Y., Lin, Z. Y., Chen, S. C., Wang, L. Y., Chang, W. Y., and Chuang, W. L. (2008). Rapid virological response and treatment duration for chronic hepatitis C genotype 1 patients: a randomized trial. *Hepatology* 47(6), 1884-93.

Yu, M. L., Huang, C. F., Huang, J. F., Chang, N. C., Yang, J. F., Lin, Z. Y., Chen, S. C., Hsieh, M. Y., Wang, L. Y., Chang, W. Y., Li, Y. N., Wu, M. S., Dai, C. Y., Juo, S. H., and Chuang, W. L. (2011). Role of interleukin-28B polymorphisms in the treatment of hepatitis C virus genotype 2 infection in Asian patients. *Hepatology* 53(1), 7-13.

Yudkin, J. S., Stehouwer, C. D., Emeis, J. J., and Coppack, S. W. (1999). C-reactive protein in healthy subjects: associations with obesity, insulin resistance, and endothelial dysfunction: a potential role for cytokines originating from adipose tissue? *Arterioscler Thromb Vasc Biol* 19(4), 972-8.

Zacho, J., Tybjaerg-Hansen, A., Jensen, J. S., Grande, P., Sillesen, H., and Nordestgaard, B. G. (2008). Genetically elevated C-reactive protein and ischemic vascular disease. *N Engl J Med* 359(18), 1897-908.

Zein, C. O., Levy, C., Basu, A., and Zein, N. N. (2005). Chronic hepatitis C and type II diabetes mellitus: a prospective cross-sectional study. *Am J Gastroenterol* 100(1), 48-55.

Zethelius, B., Berglund, L., Sundstrom, J., Ingelsson, E., Basu, S., Larsson, A., Venge, P., and Arnlov, J. (2008). Use of multiple biomarkers to improve the prediction of death from cardiovascular causes. *N Engl J Med* 358(20), 2107-16.

Zignego, A. L., Ferri, C., Giannini, C., Monti, M., La Civita, L., Careccia, G., Longombardo, G., Lombardini, F., Bombardieri, S., and Gentilini, P. (1996). Hepatitis C virus genotype analysis in patients with type II mixed cryoglobulinemia. *Ann Intern Med* 124(1 Pt 1), 31-4.

4

Influenza C Virus: Structure and Function of M Gene and Its Products

Yasushi Muraki
Department of Microbiology
Kanazawa Medical University School of Medicine
Japan

1. Introduction

Influenza C virus (Fig.1), which belongs to the genus *Influenza C Virus* of the family *Orthomyxoviridae,* was first isolated from a patient with respiratory illness in 1947 (Taylor, 1949). It is widely distributed throughout the world and the majority of humans acquire antibodies to the virus early in life (Homma et al., 1982; Nishimura et al., 1987). The virus usually causes a mild upper respiratory illness (Katagiri et al., 1983), but can also cause lower respiratory infections such as bronchitis and pneumonia (Moriuchi et al., 1991; Matsuzaki et al., 2006). Recently, a case of acute encephalopathy associated with influenza C virus infection has been reported for the first time (Takayanagi et al., 2009). Although influenza C virus is isolated infrequently due to lack of facilities equipped with the resources for performing efficient virus isolation, recurrent infection with this virus occurs frequently in children as well as in adults (Homma et al., 1982; Katagiri et al., 1983, 1987; Matsuzaki et al., 1990).

The genome of influenza C virus consists of seven RNA segments of negative polarity, each of which encodes three polymerase proteins (PB2, PB1, and P3), hemagglutinin-esterase-fusion (HEF) glycoprotein, nucleoprotein (NP), matrix (M1) protein and CM2, and two nonstructural proteins (NS1 and NS2/NEP) (Palese & Shaw, 2007). PB2, PB1 and P3 are subunits of the RNA polymerase of the virus (Crescenzo-Chaigne et al., 1999; Crescenzo-Chaigne & van der Werf, 2001; Nagele & Meier-Ewert 1984; Yamashita et al., 1989). HEF, which has receptor-binding, receptor-destroying and fusion activities, forms a spike on the virion (Herrler & Klenk, 1991). NP participates in forming ribonucleoproteins (RNPs) with viral RNA (vRNA), PB2, PB1 and P3 (Crescenzo-Chaigne et al., 1999; Crescenzo-Chaigne & van der Werf, 2001; Nakada et al., 1984; Sugawara et al., 1991; 2006). M1 is abundantly present beneath the envelope, which gives rigidity to the virion. CM2 is the second membrane protein of the virus. NS1 is involved in viral mRNA splicing (Muraki et al. 2010), and NS2/NEP is a nuclear export protein (Paragas et al., 2001) and is incorporated into the virions (Kohno et al. 2009).

In this chapter, the author will focus on the M gene and M gene products and describe how they contribute to the replication of influenza C virus. First, the author will show the M gene structure, including its coding strategy. Second, the author will mention the characteristics of M1 and CM2 proteins. Finally, the author will describe recent findings on M1 and CM2 proteins with regard to their roles in virus replication.

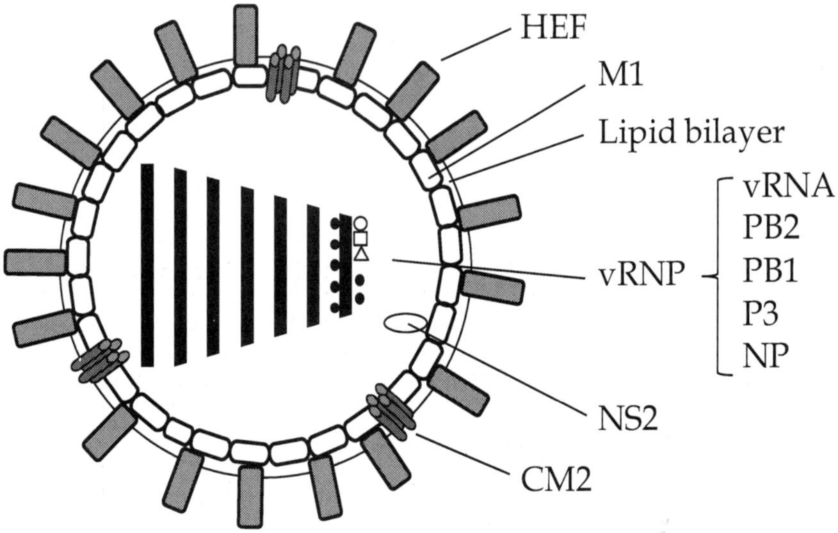

Fig. 1. Structure of influenza C virus.
The influenza C virus contains seven single-stranded RNA segments of negative polarity. The viral ribonucleoprotein (vRNP) complex is composed of viral RNA (black line), PB2 (open circle), PB1 (open rectangle), PA (open triangle) and NP (closed circle) proteins. The structure of vRNP is depicted only for the RNA Segment 7 for simplicity. HEF is a spike protein of the virus, forming homotrimers on the envelope. M1 is located beneath the envelope. CM2 is the second membrane protein that is present as a homotetramer. A small amount of NS2/NEP is incorporated into the virions.

2. Coding strategy of influenza C virus M gene

The influenza C virus RNA segment 6 (M gene) is 1,180 nucleotides in length and encodes a 242-amino-acid matrix protein (M1) and 115-amino-acid CM2 protein. In this section, the author will describe the coding strategy of the M gene, including the brief history of M1 and CM2 discovery.

2.1 M1 protein

Yamashita et al. determined the nucleotide sequence of C/Ann Arbor/1/50 M gene and provided evidence that a 242-amino-acid M1 protein is encoded by a spliced mRNA of the gene (Yamashita et al., 1988). The spliced mRNA of the gene lacks a region from nucleotides 754 to 981 due to the introduction of the stop codon (TGA) as a result of splicing (Fig. 2). The splicing event itself introduced the stop codon TGA, with the TG coming from the 5′ splicing site and the A coming from the 3′ splicing site.

The mRNA for M1 represents the major M gene-specific mRNA species in virus-infected cells. It is of interest that the splicing pattern of the influenza C virus M gene is different from that of influenza A virus M gene, in that the influenza A virus M1 protein, a major product of the gene, is coded for by an unspliced mRNA, not spliced mRNA (Palese & Shaw, 2007).

An open reading frame (ORF) in the influenza C virus M gene could potentially code for a protein of 374 amino acids, and small quantities of an mRNA collinear with RNA segment 6 was actually detected since a faint signal corresponding to the mRNA species was observed by S1 analysis (Yamashita et al., 1988). At that moment, however, the significance of the collinear mRNA species remained unknown.

2.2 CM2 protein

Analysis of the nucleotide sequences of the M gene from five influenza C virus isolates, including C/Ann Arbor/1/50, revealed that the gene contains a single ORF capable of coding for a 374-amino-acid protein (Hongo et al., 1994; Yamashita et al., 1988). Therefore, attempts was made to identify the unspliced mRNA of the M gene, and the mRNA was found to be synthesized in 1/10 amount of the spliced mRNA in the virus-infected cells (Hongo et al., 1994). To identify the protein encoded by the unspliced mRNA, whose approximate molecular weight is 42,000, the antiserum against the glutathione S-transferase fusion protein containing the extra C-terminal domain of the protein (Fig. 2) was prepared. Unexpectedly, immunoprecipitation experiments with the antiserum identified a protein of Mr ~18,000 in virus-infected cells (Hongo et al., 1994). The authors designated the protein as CM2, a second protein encoded by the influenza C virus M gene, and identified three forms of CM2 (CM2o, CM2a, and CM2b) depending on the moieties of glycosylation (Hongo et al., 1994).

A previously unrecognized 374-amino-acid protein was identified in virus-infected cells using the antiserum against CM2, and the protein was named as P42 according to the molecular weight. P42 is demonstrated to be modified by the addition of a high-mannose oligosaccharide chain to generate P44. The tryptic peptide map of either P42 or P44 was indistinguishable from the map of the mixture of M1 and CM2. Thus, the authors concluded that P42 and P44 correspond to the 374 amino acid protein encoded by unspliced RNA segment 6 mRNA and its N-glycosylated form, respectively (Hongo et al., 1998).

At that moment, the translational mechanism of CM2 was still unclear. It was reported previously that the P42 is an integral membrane protein having two internal hydrophobic domains, one of which (residues 241 to 252) is followed by two sequences (252 Ile-Thr-Ser and 257 Ala-Ser-Ala) favourable for cleavage by signal peptidase. To examine the possibility that P42 is cleaved by signal peptidase to yield CM2, a series of mutant M gene cDNAs were constructed and transfected into cultured cells (Hongo et al., 1999; Pekosz & Lamb, 1998). As a result, biosynthesis of CM2 has been demonstrated to proceed in the following manner; (i) Unspliced mRNA from RNA segment 6 is first translated into a 374-amino-acid protein, P42. (ii) Cotranslationally or immediately after the completion of translation, P42 begins to be inserted into the endoplasmic reticulum with the aid of the first hydrophobic domain, composed of amino acid residues 241 to 252. (iii) Translocation of P42 through the membrane is halted by the presence of the second hydrophobic domain, consisting of residues 287 to 318. (iv) P42 is then N-glycosylated at Asn residue 270, generating P44. (v)

Either before or after the addition of an oligosaccharide chain, P42/P44 is cleaved by signal peptidase at the C-terminal side of Ala residue 259, producing the M1′ and CM2 proteins, composed of the N-terminal 259 amino acids and the C-terminal 115 amino acids, respectively.

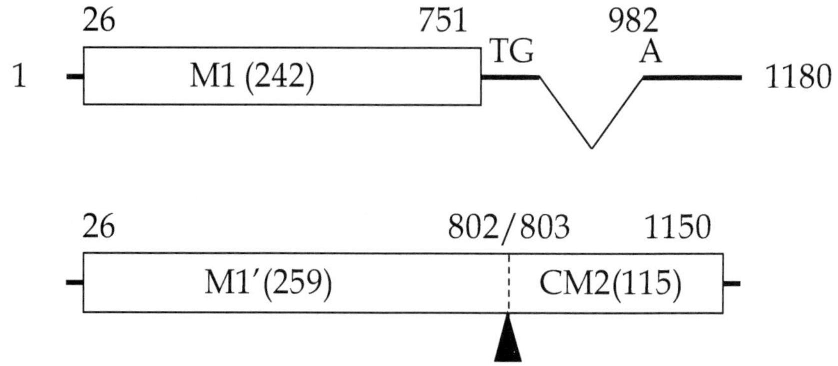

Fig. 2. Coding strategy of influenza C virus M gene.
RNA segments 6 (M gene) of C/Ann Arbor/1/50 is shown in positive-sense orientation. Numbers indicate the nucleotide positions along the genes. The lines at the 5′ and 3′ termini represent the non-coding regions. The boxes represent the coding region of viral proteins encoded by the gene. The intron is shown by V-shaped lines. M1 is encoded by a spliced mRNA (upper panel), into which the stop codon TGA is introduced at 752, 753 and 982, as a result of splicing. The P42 protein (M1′+CM2), encoded by a collinear transcript of the gene (lower panel), is cleaved by signal peptidase at an internal cleavage site (closed triangle) to generate M1′ and CM2.

3. Characteristics of the M1 and CM2 proteins

3.1 M1 protein

The influenza C virion contains three major structural polypeptides: a large glycoprotein gp88 (HEF), a nucleoprotein (NP) and matrix protein (M1) (Palese & Shaw, 2007). Furthermore, Yokota et al. identified the M1 protein synthesized in C/Ann Arbor/1/50-infected MDCK cells (Yokota et al., 1981). To characterize the protein, the monoclonal antibodies against M1 have been produced (Sugawara et al., 1991). Using a panel of the nine MAbs against M1, the M1 protein was shown to contain two non-overlapping antigenic regions that are highly resistant to conformational changes, and to exhibit no antigenic variations among 23 influenza C virus strains isolated over 41 year period. The C/Ann Arbor/1/50-infected cells were examined for the localization of the M1 protein using one of these MAbs (L2), and as a result, M1 is shown to be localized in the nucleus at 24 h postinfection. The latter finding provided evidence for the requirement of nuclear function for influenza C virus replication. In addition, the association of M1 with the nucleoli was observed in the C/Yamagata/1/88-infected cells, a phenomenon that remains to be elucidated with respect to viral replication (Sugawara et al., 1991).

3.2 CM2 protein

The biochemical characteristics of CM2 in virus-infected cells were precisely determined, and as a result, CM2 is demonstrated to be the counterpart of the influenza A virus M2 protein (Hongo et al., 1997; Pekosz & Lamb, 1997; Tada et al., 1998). CM2 is a type III integral membrane protein that is oriented in membranes with a 23-amino-acid N-terminal extracellular domain, a 23-amino-acid transmembrane domain and a 69-amino-acid C-terminal cytoplasmic domain. It forms homodimer and homotetramer, and is phosphorylated, palmitoylated and N-glycosylated, and is incorporated into progeny virions. Site-specific mutational analyses of the M gene identified the amino acid sites for the posttranslational modifications on CM2 molecules as follows; i) an asparagine residue 11 is an N-glycosylation site, ii) cysteines at residue 1, 6 and 20 are all involved in disulfide bond formation, iii) a cysteine residue at 65 is palmitoylated through labile thioester linkage, and iv) serine residues at 78, 103, 108 and a proline at 104 are phosphorylation sites (Li et al., 2001; Pekosz & Lamb, 1997).

CM2 has also been investigated from an electrophysiological point of view. A number of studies have shown that the CM2 protein appears to have ion channel activities. CM2 forms a voltage-activated ion channel permeable to Cl^- (Hongo et al., 2004). When co-expressed with a pH-sensitive hemagglutinin (HA) from influenza A virus, the CM2 protein has a capacity to reduce the acidity of the exocytic pathway and reduce conversion of the pH-sensitive HA to its low pH conformation during transport to the cell surface, suggesting that CM2 is permeable to H^+ (Betáková & Hay, 2007). Preliminary electrophysiological studies of CM2-expressing mouse erythroleukemia cells have identified Na^+-activated proton permeability in addition to the low pH-activated Cl^- permeability (Muraki & Hay, 2009; Muraki et al., personal communications). However, the relationship between the channel function and the role(s) of CM2 in the virus replication (see below) remains to be clarified.

4. Role of the M gene-products in virus replication

4.1 M1 protein

The M1 protein of influenza C virus is involved in virion morphogenesis. Nishimura et al. reported that cord-like structures (CLSs) which had lengths up to ~500 μm or greater were protruding from the surface of C/Yamagata/1/88-infected HMV-II cells (Nishimura et al., 1990). Electron microscopic analysis showed that CLS consists of numerous filamentous particles in the process of budding, each of which is covered with a layer of surface projections and aggregated with their long axes. Further analyses using a series of reassortant viruses between C/Yamagata/1/88 and C/Taylor/1233/47, the latter of which is a unique strain incapable of forming CLS, showed that reassortants with the M gene from C/Taylor/1233/47 could not form CLS on infected cells. Comparison of the M gene sequences of influenza C virus strains, including C/Yamagata/1/88 and C/Taylor/1233/47, suggested that either or both of amino acid changes at positions 24 and 133 of the M1 protein were responsible for differences in cord-forming ability between the two strains (Nishimura et al., 1994).

The amino acid on the M1 protein that is responsible for CLS formation and virion morphology was determined. Upon establishment of influenza C virus-like particle (VLP) generation system, Muraki et al. observed that CLS was protruding from the 293T cells transfected with a set of plasmid DNAs for VLP generation (Muraki et al., 2004). CLSs were observed on the 293T cells transfected with the plasmid for M1 of C/Ann Arbor/1/50, whereas CLSs were not observed when the plasmid for M1 of C/Ann Arbor/1/50 was replaced with that of C/Taylor/1233/47. Expression of the M1 protein possessing alanine or threonine at residue 24 together with the other virus components resulted in the generation of filamentous or spherical VLPs, respectively. These findings indicate that an amino acid at residue 24 of the M1 protein is a determinant for virion morphology.

The above finding obtained from VLP generation system has been confirmed using a recombinant virus generated by reverse genetics. The parental recombinant influenza C virus, which has the consensus sequences of C/Ann Arbor/1/50 (the residue 24 of M1 is alanine), exhibited a filamentous morphology. In contrast, a recombinant mutant virus, rMG96A, which has an Ala→Thr mutation at residue 24 of M1, exhibited spherical morphology. The mutant M1 protein had lower membrane affinity than did the wild type M1, suggesting that the difference in the affinity affects the virion morphology (Muraki et al., 2007). Furthermore, the flotation analysis showed that budding of influenza C virus does not occur from the lipid raft domain of the plasma membrane. These studies indicate that M1 plays an important role in virus budding, and imply that a region around the residue 24 of M1 may represent a late domain of the protein and that budding mechanism of influenza C virus differs from that of influenza A virus.

4.2 CM2 protein

The fact that the biochemical characteristics of CM2 are closely similar to those of influenza A virus M2 and that CM2 seems to have permeability to proton suggests that CM2 is also crucial in influenza C virus replication. To clarify the role(s) of CM2 in the influenza C virus replication cycle, attempts were made to generate a recombinant CM2-deficient virus using reverse genetics. However, infectious recombinants lacking CM2 have not been rescued to date (Furukawa et al., 2011), which highly suggests that CM2 is essential to influenza C virus replication.

Using VLP generation system, evidence was obtained that CM2 is involved in the packaging and uncoating processes. The CM2-deficient influenza C VLPs were successfully generated and the amount of GFP-vRNA in the VLPs was quantified. The data showed that the CM2-deficient VLPs contain approximately 37% of the vRNA found in wild-type VLPs, although no significant differences were detected in the expression levels of viral components in VLP-producing cells as well as in the number and morphology of the generated VLPs (Furukawa et al., 2011). This finding suggests that CM2 is involved in the mini-genome packaging into VLPs. In addition, HMV-II cells infected with CM2-deficient VLPs exhibited significantly reduced GFP expression. Although CM2-deficient VLPs could be internalized into HMV-II cells as efficiently as wild-type VLPs, a smaller amount of GFP-vRNA was detected in the nuclear fraction of CM2-deficient VLP-infected cells than in that of wild-type VLP-infected cells, suggesting that the uncoating process of the CM2-deficient VLPs in the infected cells did not proceed in an appropriate manner (Furukawa et al., 2011). Thus, CM2 appears to play a role in the packaging and uncoating processes of the virus replication cycle. However, whether and how the CM2 channel activities relate to the roles remains unclear.

A recombinant influenza C virus lacking CM2 palmitoylation, rCM2-C65A, was successfully rescued (Muraki et al., 2011). The rCM2-C65A, in which the cysteine at residue 65 of CM2 was substituted to alanine, grew as efficiently as did the parental recombinant virus in cultured cells and embryonated chicken eggs. The transport, maturation and localization of HEF, NP, M1 and CM2 in the cells infected with rCM2-C65A were virtually identical to those of the parental virus-infected cells. These findings suggest that CM2 palmitoylation is not required for virus replication. The fact that a recombinant lacking CM2 palmitoylation could be rescued suggests that CM2 mutant(s) at posttranslational site(s) is available for the analysis of CM2 role(s) in the virus replication.

The ion channel proteins of influenza A and B viruses play critical roles in the viral replication. The influenza A virus M2 protein functions as a proton channel during the entry of the virus into cells by allowing the acidification of the virion interior (Palese & Shaw, 2007). Recently, the M2 proton-selective ion channel function mediates virus budding (Rossman et al., 2010b; 2011). The M2 cytoplasmic tail also plays a role in progeny virus production and virion morphology (Chen et al., 2008; Iwatsuki-Horimoto et al. 2006; McCown & Pekosz, 2005; Rossman et al., 2010a; Zhang et al., 2000). The influenza B virus BM2 protein has a proton channel activity (Mould et al., 2003), and is involved in the incorporation of the virus genome into virions (Imai et al., 2004; 2008). Thus CM2 needs to be further investigated from these points of view.

4.3 M1' and P42

As mentioned above, P42, encoded by a collinear transcript of the RNA segment 6 (M gene), is cleaved by signal peptidase to generate CM2, composed of the C-terminal 115 amino acids, in addition to the M1' protein, composed of the N-terminal 259 amino acids (Hongo S et al., 1999; Pekosz & Lamb 1998). It is of interest to clarify the role(s) of P42 and M1' in virus replication. The characteristics of the P42 and M1' (or p31) proteins were precisely determined as follow.

The M1' protein was found to undergo rapid degradation after cleavage from P42, since addition of the 26S proteasome inhibitor lactacystin to the cells infected with influenza C virus or transfected with M gene cDNA drastically reduced M1' degradation (Pekosz & Lamb 2000). In addition, the hydrophobic nature, not the specific amino acid sequence, of the 17-amino-acid C terminus of M1' was demonstrated to act as the signal for targeting the protein to membranes and for degradation (Pekosz & Lamb 2000). Thus, M1' seems to be dispensable to virus replication.

A series of mutant M gene cDNAs were transfected into COS-1 cells to examine the characteristics of P42 (Li et al., 2004). P44, the N-glycosylated form of P42, forms disulfide-linked dimers and tetramers. P44 is transported to the Golgi apparatus, but not to the trans-Golgi, since P44 is sensitive to endoglycosidase H. P44 and P42 are unstable irrespective of N-glycosylation or oligomerization. The addition of lactacystin prevented the degradation of P42 as well as M1', but not that of P44 efficiently, suggesting that P44 is degraded by another protease besides the 26S proteasome (Li et al., 2004). The role(s) of P42/P44 in virus replication remains to be clarified.

5. Regulation of M1 and CM2 synthesis

In influenza A virus-infected cells, the regulatory machanism for viral mRNA splicing has been extensively investigated. It has been reported that the steady-state level of spliced viral transcripts is only 10% that of unspliced viral transcripts (Palese & Shaw, 2007). The inefficient splicing can be understood partly by the fact that influenza A virus NS1 protein is associated with spliceosomes and inhibits pre-mRNA splicing (Fortes et al., 1994; Garaigorta & Ortin, 2007; Lu et al., 1994). The cis-acting sequences in the NS1 transcript also negatively regulates splicing (Nemeroff et al., 1992). The splicing of influenza A virus M1 mRNA is controlled by the rate of nuclear export (Valcarcel et al., 1993), and the splicing of influenza A virus M1 mRNA is regulated by the binding of the viral polymerase complex and cellular splicing factor SF2/ASF (Shih et al., 1995 ; Shih & Krug, 1996).

The influenza C virus NS1 protein (C/NS1), a product of the RNA segment 7 (NS gene), regulates the splicing efficiency of the M gene (Muraki et al., 2010). In influenza C virus-infected cells, the predominant transcript from the M gene is a spliced mRNA (Hongo et al., 1994). Ribonuclease protection assays showed that the splicing efficiency of the M gene mRNA in infected cells was much higher than that in M gene cDNA-transfected cells, suggesting that viral factor(s) facilitates the M gene mRNA splicing. To investigate the factor(s) involved in the splicing, Muraki et al. analyzed cells co-transfected with M and NS gene-cDNAs, and showed that the splicing of the M gene mRNA was enhanced by expressing C/NS1 (Muraki et al., 2010). The fact was obtained that splicing ratio of influenza A virus M gene mRNA was also increased by C/NS1, suggesting that C/NS1 has an intrinsic ability of up-regulating viral mRNA splicing. Therefore, we speculate that efficient splicing of M gene mRNA leads to a high and reduced level of M1 and CM2 expression, respectively, creating conditions that are optimal for virus replication.

6. Conclusion

The RNA segment 6 (M gene) of influenza C virus encodes the M1 and CM2 proteins. It has been demonstrated that the M1 protein is a determinant for virion morphology and CM2 is involved in packaging and uncoating processes. A further study will be needed to clarify the structure-function relationship more precicely.

7. Acknowledgements

The author thanks all the laboratory members of Department of Microbiology, Kanazawa Medical University School of Medicine and Department of Infectious Diseases, Yamagata University Faculty of Medicine for helpful assistance and discussion.

8. References

Betáková, T. & Hay, A.J. (2007). Evidence that the CM2 Protein of Influenza C Virus Can Modify the pH of the Exocytic Pathway of Transfected Cells. *Journal of General Virology*, Vol.88, pp.2291-2296, ISSN 1465-2099

Chen, B.J.; Leser, G.P.; Jackson, D. & Lamb, R.A. (2008). The Influenza Virus M2 Protein Cytoplasmic Tail Interacts with the M1 Protein and Influences Virus Assembly at the Site of Virus Budding. *Journal of Virology*, Vol.82, No.20, pp.10059-10070, ISSN 0022-538X

Crescenzo-Chaigne, B.; Naffakh, N. & van der Werf, S. (1999). Comparative Analysis of the Ability of the Polymerase Complexes of Influenza Viruses Type A, B and C to Assemble into Functional RNPs that Allow Expression and Replication of

Heterotypic Model RNA Templates *In Vivo*. *Virology*, Vol.265, pp.342-353, ISSN 0042-6822

Crescenzo-Chaigne, B. & van der Werf, S. (2001). Nucleotides at the Extremities of the Viral RNA of Influenza C Virus are Involved in Type-Specific Interactions with the Polymerase Complex. *Journal of General Virology*, Vol.82, pp.1075-1083, ISSN 1465-2099

Fortes, P.; Beloso, A. & Ortin, J. (1994). Influenza Virus NS1 Protein Inhibits Pre-mRNA Splicing and Blocks mRNA Nucleocytoplasmic Transport. *EMBO Journal*, Vol.13, pp.704–712, ISSN 0261-4189

Furukawa, T.; Muraki, Y.; Noda, T.; Takashita, E.; Sho, R.; Sugawara, K.; Matsuzaki, Y.; Shimotai, Y. & Hongo, S. (2011). Role of the CM2 Protein in the Influenza C Virus Replication Cycle. *Journal of Virology*,Vol.85, No.3, pp.1322–1329, ISSN 0022-538X

Garaigorta, U. & J. Ortin, J. (2007). Mutation Analysis of a Recombinant NS Replicon Shows that Influenza Virus NS1 Protein Blocks the Splicing and Nucleo-Cytoplasmic Transport of its own Viral mRNA. *Nucleic Acids Research*, Vol.35, pp.4573-4582, ISSN 0305-1048

Herrler, G. & Klenk, H.D. (1991). Structure and Function of the HEF Glycoprotein of Influenza C Virus, In: *Advances in Virus Research*, Maramorosch, K.; Frederick A. Murphy, F.A. & ShatkinA.J., Vol.40, pp.213–234, ISBN 978-0-12-374322-0, Elsevier, Amsterdam, The Netherland

Homma, M.; Ohyama, S. & Katagiri, S. (1982). Age Distribution of the Antibody to Type C Influenza Virus. *Microbiology and Immunology*, Vol.26, pp.639–642, ISSN 0385-5600

Hongo, S.; Sugawara, K.; Nishimura, H.; Muraki, Y.; Kitame, F. & Nakamura, K. (1994). Identification of a Second Protein Encoded by Influenza C Virus RNA Segment 6. *Journal of General Virology*, Vol.75, pp.3503–3510, ISSN 0022-1317

Hongo, S.; Sugawara, K.; Muraki, Y.; Kitame, F. & Nakamura, K. (1997). Characterization of a Second Protein (CM2) Encoded by RNA Segment 6 of Influenza C Virus. *Journal of Virology*, Vol.71, No.4, pp.2786–2792, ISSN 0022-538X

Hongo, S.; Gao, P.; Sugawara, K.; Muraki, Y.; Matsuzaki, Y.; Tada, Y.; Kitame, F. & Nakamura, K. (1998). Identification of a 374 Amino Acid Protein Encoded by RNA Segment 6 of Influenza C Virus. *Journal of General Virology*, Vol.79, pp.2207-2213, ISSN 0022-1317

Hongo, S.; Sugawara, K.; Muraki, Y.; Matsuzaki, Y.; Takashita, E.; Kitame, F. & Nakamura, K. (1999). Influenza C Virus CM2 Protein is Produced From a 374-Amino-Acid Protein (P42) by Signal Peptidase Cleavage. *Journal of Virology*, Vol.73, No.1, pp.46–50, ISSN 0022-538X

Hongo, S.; Ishii, K.; Mori, K.; Takashita, E.; Muraki, Y.; Matsuzaki, Y. & Sugawara, K. (2004). Detection of Ion Channel Activity in *Xenopus Laevis* Oocytes Expressing Influenza C Virus CM2 Protein. *Archives of Virology*, Vol.149, pp.35–50, ISSN 0304-8608

Imai, M.; Watanabe, S.; Ninomiya, A.; Obuchi, M. & Odagiri, T. (2004). Influenza B Virus BM2 Protein is a Crucial Component for Incorporation of Viral Ribonucleoprotein Complex into Virions during Virus Assembly. *Journal of Virology*, Vol.78, pp.10007-10015, ISSN 0022-538X

Imai, M.; Kawasaki, K. & Odagiri, T. (2008). Cytoplasmic Domain of Influenza B Virus BM2 Protein Plays Critical Roles in Production of Infectious Virus. *Journal of Virology*, Vol.82, pp.728-739, ISSN 0022-538X

Iwatsuki-Horimoto, K.; Horimoto, T.; Noda, T.; Kiso, M.; Maeda, J.; Watanabe, S.; Muramoto, Y.; Fujii, K. & Kawaoka, Y. (2006). The Cytoplasmic Tail of the

Influenza A Virus M2 Protein Plays a Role in Viral Assembly. *Journal of Virology*,Vol.80, pp.5233–5240, ISSN 0022-538X

Katagiri, S.; Ohizumi, A. & Homma, M. (1983). An Outbreak of Type C Influenza in a Children's Home. *Journal of Infectious Diseases,* Vol.148, pp.51–56, ISSN 0022-1899

Katagiri, S.; Ohizumi, A.; Ohyama, S. & Homma, M. (1987). Follow-Up Study of Type C Influenza Outbreak an a Children's Home. *Microbiology and Immunology,* Vol.31, pp.337–343, ISSN 0385-5600

Kohno, Y.; Muraki, Y.; Matsuzaki, Y.; Takashita, E.; Sugawara, K. & Hongo, S. (2009). Intracellular Localization of Influenza C Virus NS2 Protein (NEP) in Infected Cells and its Incorporation into Virions. *Archives of Virology,* Vol.154, pp.235–243, ISSN 0304-8608

Li, Z.N.; Hongo, S.; Sugawara, K.; Tsuchiya, E.; Matsuzaki, Y. & Nakamura, K. (2001). The Sites for Fatty Acylation, Phosphorylation and Intermolecular Disulphide Bond Formation of Influenza C Virus CM2 Protein. *Journal of General Virology,* Vol.82, pp.1085–1093, ISSN 0022-1317

Li, Z.N.; Muraki, Y.; Takashita, E.; Matsuzaki, Y.; Sugawara, K. & Hongo, S. (2004). Biochemical Properties of the P42 Protein Encoded by RNA Segment 6 of Influenza C Virus. *Archives of Virology,* Vol.149, pp.275–287, ISSN 0304-8608

Lu, Y.; Qian, X.Y. & Krug, R.M. (1994). The Influenza Virus NS1 Protein: A Novel Inhibitor of Pre-mRNA Splicing. *Genes & Developement,* Vol.8, pp.1817-1828, ISSN 0890-9369

Matsuzaki, M.; Adachi, K.; Sugawara, K.; Nishimura, H.; Kitame, F. & Nakamura, K. (1990). A Laboratory-Acquired Infection with Influenza C Virus. *Yamagata Medical Journal,* Vol.8, pp.41–51, ISSN 0288-030X

Matsuzaki, Y.; Katsushima, N.; Nagai, Y.; Shoji, M.; Itagaki, T.; Sakamoto, M.; Kitaoka, S.; Mizuta, K. & Nishimura, H. (2006). Clinical Features of Influenza C Virus Infection in Children. *Journal of Infectious Diseases,* Vol.193, pp.1229–1235, ISSN 0022-1899

McCown, M.F. & Pekosz, A. (2005). The Influenza A Virus M2 Cytoplasmic Tail is Required for Infectious Virus Production and Efficient Genome Packaging. *Journal of Virology,* Vol.79, pp.3595-3605, ISSN 0022-538X

Moriuchi, H.; Katsushima, N.; Nishimura, H.; Nakamura, K. & Numazaki, Y. (1991). Community-Acquired Influenza C Virus Infection in Children. *Journal of Pediatrics,* Vol.118, pp.235–238, ISSN 0022-3476

Mould, J.A.; Paterson, R.G.; Takeda, M.; Ohigashi, Y.; Venkataraman, P.; Lamb, R.A. & Pinto, L.H. (2003). Influenza B Virus BM2 Protein has Ion Channel Activity that Conducts Protons across Membranes. *Developmental Cell,* Vol.5, pp.175-184, ISSN 1534-5807

Muraki, Y.; Washioka, H.; Sugawara, K.; Matsuzaki, Y.; Takashita, E. & Hongo, S. (2004). Identification of an Amino Acid Residue on Influenza C Virus M1 Protein Responsible for Formation of the Cord-Like Structures of the Virus. *Journal of General Virology,* Vol.85, pp.1885–1893, ISSN 0022-1317

Muraki, Y.; Murata, T.; Takashita, E.; Matsuzaki, Y.; Sugawara, K. & Hongo, S. (2007). A Mutation on Influenza C Virus M1 Protein Affects Virion Morphology by Altering the Membrane Affinity of the Protein. *Journal of Virology,*Vol.81, No.16, pp.8766–8773, ISSN 0022-538X

Muraki, Y. & Hay, A. (2009). Establishment of Mouse Erythroleukemia Cell Line Expressing Influenza C Virus CM2 Protein and A Chimeric Protein between CM2 and Influenza A Virus M2 Protein. *Acta Virologica,* Vol.53, pp.125-129, ISSN 0001-723X

Muraki, Y.; Furukawa, T.; Kohno, Y.; Matsuzaki, Y.; Takashita, E.; Sugawara, K. & Hongo, S. (2010). Influenza C Virus NS1 Protein Up-Regulates the Splicing of Viral mRNAs. *Journal of Virology,* Vol.84, No.4, pp.1957–1966, ISSN 0022-538X

Muraki, Y.; Okuwa, T.; Furukawa, T.; Matsuzaki, Y.; Sugawara, K.; Himeda, T.; Hongo, S. & Ohara, Y. (2011). Palmitoylation of CM2 is Dispensable to Influenza C Virus Replication. *Virus Research,* Vol.157, pp.99–105, ISSN 0168-1702

Nagele, A. & Meier-Ewert, H. (1984). Influenza-C-Virion-Associated RNA-Dependent RNA-Polymerase Activity. *Bioscience Reports,* Vol.4, pp.703-706, ISBN 0144-8463

Nakada, S.; Creager, R.S.; Krystal, M. & Palese, P. (1984). Complete Nucleotide Sequence of the Influenza C/California/78 Virus Nucleoprotein Gene. *Virus Research,* Vol.1, pp.433-441, ISSN 0168-1702

Nemeroff, M.E.; Utans, U.; Kramer, A. & Krug, R.M. (1992). Identification of Cis-Acting Intron and Exon Regions in Influenza Virus NS1 mRNA that Inhibit Splicing and Cause the Formation of Aberrantly Sedimenting Presplicing Complexes. *Molecular and Cellular Biology*, Vol.12, pp.962-970, ISSN 0270-7306

Nishimura, H.; Sugawara, K.; Kitame, F.; K.; Nakamura, K. & Sasaki, H. (1987). Prevalence of the Antibody to Influenza C Virus in a Northern Luzon Highland Village, Philippines. *Microbiology and Immunology,* Vol.31, pp.1137–1143, ISSN 0385-5600

Nishimura, H.; Hara, M.; Sugawara, K.; Kitame, F.; Takiguchi, K.; Umetsu, Y.; Tonosaki, A. & Nakamura, K. (1990). Characterization of the Cord-Like Structures Emerging from the Surface of Influenza C Virus-Infected Cells. *Virology,* Vol.179, pp.179–188, ISSN 0042-6822

Nishimura, H.; Hongo, S.; Sugawara, K.; Muraki, Y.; Kitame, F.; Washioka, H.; Tonosaki, A. & Nakamura K. (1994). The Ability of Influenza C Virus to Generate Cord-Like Structures is Influenced by the Gene Coding for M-Protein. *Virology,* Vol.200, pp.140–147, ISSN 0042-6822

Palese, P. & Shaw, M.L. (2007). Orthomyxoviridae: the viruses and their replication, In: *Fields Virology, 5th ed.,* Knipe, D.M.; Howley, P.M.; Griffin, D.E.; Lamb, R.A.; Martin, M.A.; Roizman, B. & Straus S.E., pp.1647–1689, Lippincott Williams & Wilkins, ISBN 978-0-7817-6060-7, Philadelphia, PA

Paragas, J.; Talon, J.; O'Neill, R.E.; Anderson D.K.; García-Sastre, A. & Palese, P. (2001). Influenza B and C Virus NEP (NS2) Proteins Possess Nuclear Export Activities. *Journal of Virology,* Vol.75, pp. 7375–7383, ISSN 0022-538X

Pekosz, A. & Lamb, R.A. (1997). The CM2 Protein of Influenza C Virus is an Oligomeric Integral Membrane Glycoprotein Structurally Analogous to Influenza A Virus M2 and Influenza B Virus NB Proteins. *Virology,* Vol.237, pp.439–451, ISSN 0042-6822

Pekosz, A. & Lamb, R.A. (1998). Influenza C Virus CM2 Integral Membrane Glycoprotein is Produced from a Polypeptide Precursor by Cleavage of an Internal Signal Sequence. *Proceedings of National Academy of Science U.S.A.,* Vol.95, pp. 13233 - 13238, ISSN 1091-6490

Pekosz, A. & Lamb, R.A. (2000). Identification of a Membrane Targeting and Degradation Signal in the P42 Protein of Influenza C Virus. *Journal of Virology,*Vol.74, pp.10480–10488, ISSN 0022-538X

Rossman, J.S.; Jing, X.; Leser, G.P.; Balannik, V.; Pinto, L.H. & Lamb, R.A. (2010a). Influenza Virus M2 Ion Channel Protein is Necessary for Filamentous Virion Formation. *Journal of Virology* , Vol.84, pp.5078–5088, ISSN 0022-538X

Rossman, J.S.; Jing, X.; Leser, G.P. & Lamb, R.A. (2010b). Influenza Virus M2 Protein Mediates ESCRT-Independent Membrane Scission. *Cell*, Vol.142, No.6, pp.902–913, ISSN 0092-8674

Rossman, J.S. & Lamb, R.A. (2011). Influenza Virus Assembly and Budding. *Virology*, Vol.411, pp.229-236, ISSN 0042-6822

Shih, S.R.; Nemeroff, M.E. & Krug, R.M. (1995). The Choice of Alternative 5′ Splice Sites in Influenza Virus M1 mRNA is Regulated by the Viral Polymerase Complex. *Proceedings of National Academy of Science U.S.A.*, Vol.92, pp.6324-6328, ISSN 1091-6490

Shih, S.R. & Krug, R.M. (1996). Novel Exploitation of a Nuclear Function by Influenza Virus: The Cellular SF2/ASF Splicing Factor Controls the Amount of the Essential Viral M2 Ion Channel Protein in Infected Cells. *EMBO Journal*, Vol.15, pp.5415-5427, ISSN 0261-4189.

Sugawara, K.; Nishimura, H.; Hongo, S.; Kitame, F. & Nakamura, K. (1991). Antigenic Characterization of the Nucleoprotein and Matrix Protein of Influenza C Virus with Monoclonal Antibodies. *Journal of General Virology*, Vol.72, pp.103–109, ISSN 0022-1317

Sugawara, K.; Muraki, Y.; Takashita, E.; Matsuzaki, Y. & Hongo, S. (2006). Conformational Maturation of the Nucleoprotein Synthesized in Influenza C Virus-Infected Cells. *Virus Research*,Vol.122, pp.45-52, ISSN 0168-1702

Tada, Y.; Hongo, S.; Muraki, Y.; Matsuzaki, Y.; Sugawara, K.; Kitame, F. & Nakamura, K. (1998). *Virus Research*, Vol.58, pp. 65–72, ISSN 0168-1702

Takayanagi, M.; Umehara, N.; Watanabe, H.; Kitamura, T.; Ohtake, M.; Nishimura, H.; Matsuzaki, Y. & Ichiyama, T. (2009). Acute Encephalopathy Associated with Influenza C Virus Infection. *Pediatric Infectious Disease Journal*, Vol.28, No.6, pp.554, ISSN 0891-3668

Taylor, R. M. (1949). Studies on Survival of Influenza Virus Between Epidemics and Antigenic Variants of the Virus. *American Journal of Public Health*, Vol.39, pp.171–178, ISSN 0090-0036

Valcarcel, J.; Fortes, P. & Ortin, J. (1993). Splicing of Influenza Virus Matrix Protein mRNA Expressed from a Simian Virus 40 Recombinant. *Journal of General Virology*, Vol.74, pp.1317-1326, ISSN 0022-1317

Yamashita, M.; Krystal, M. & Palese, P. (1988). Evidence that the Matrix Protein of Influenza C Virus is Coded for by a Spliced mRNA. *Journal of Virology*, Vol.62, No.9, pp.3348–3355, ISSN 0022-538X

Yamashita, M.; Krystal, M. & Palese, P. (1989). Comparison of the 3 Large Polymerase Proteins of Influenza A, B, and C Viruses. *Virology*, Vol.171, pp.458-466, ISSN 0042-6822

Yokota, M.; Nakamura, K.; Sugawara, K. & Homma, M. (1983). The Synthesis of Polypeptides in Influenza C Virus-Infected Cells. *Virology*, Vol.130, pp.105–117, ISSN 0042-6822

Zhang, J.; Leser, G.P.; Pekosz, A.; Lamb, R.A. (2000). The Cytoplasmic Tails of the Influenza Virus Spike Glycoproteins are Required for Normal Genome Packaging. *Virology*, Vol.269, No.2, pp.325-334, ISSN 0042-6822

5

Viral Vectors in Neurobiology: Therapeutic and Research Applications

Renata Coura
Centre de Neuroscience Paris Sud - CNPS - Université Paris Sud XI
France

1. Introduction

1.1 History and definition of viral vectors

Viruses are intracellular parasites with simple DNA or RNA genomes (Figure 1A). Three steps compose virus life cycle: infection of a host cell, replication of its genome within the host cell environment, and formation of new virions (Figure 1B). This process is often but

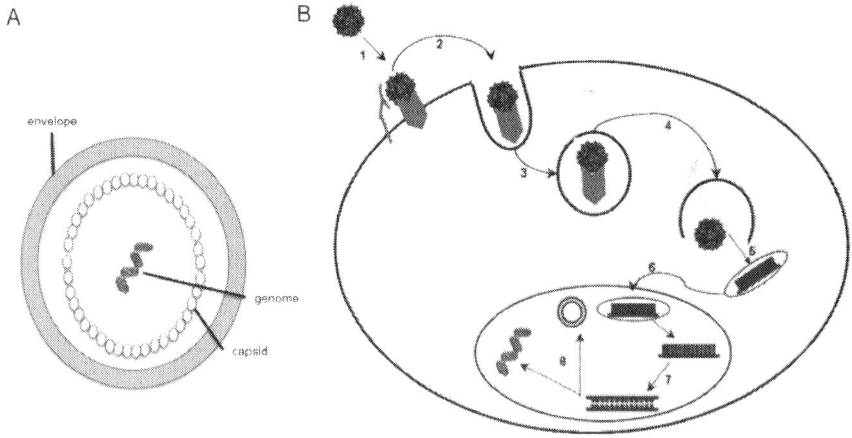

Fig. 1. Virus. (A) Structure. Simplified scheme of virus structure, with a lipid envelope that can be present or not; a protein-composed capsid and the genetic material, that can be DNA or RNA, double or single strand. (B) Life cycle. Example of the course of adeno-associated virus (AAV) productive infection. Scheme showing the eight steps of AAV transduction of host cells: (1) viral binding to a membrane receptor/co-receptor; (2) endocytosis of the virus by the host cell; (3) virus intracellular trafficking through the endosomal compartment; (4) escape of the virus from the endosome; (5) virion uncoating; (6) entry into the nucleus; (7) viral genome conversion from a single-stranded to a double-stranded genome; and (8) integration into the host genome or permanence of an episomal form capable of expressing an encoded gene (from Coura and Nardi, 2008).

not always associated with pathogenic effects against the host organism. Nevertheless, since the mid-1980s, a likely useful role for virus has been envisaged. The idea is to use the unique virus capacity to enter the cell and to replicate their genome to construct vectors, containing the viral envelope and a recombinant genome, so that these vectors could be able to deliver genetic material into cells. Then, recombinant viral vectors are created in which genes essential for viral replication are removed and a gene of interest is inserted in the viral genome (Figure 2). While this eliminates pathogenicity due to viral replication, retention of viral genes and continued expression of these genes may limit the potential of the current generation of vectors. Meanwhile, defective viral vectors represent a different approach, in which only viral recognition signals are used to allow packaging of foreign DNA into a viral coat while eliminating the possibility of viral *gene expression* (see **glossary**) within target cells. These viral vectors would be able to long-term gene delivery to mammalian cells, without pathogenicity and with minimal associated toxicity. Today, several viral vector systems are close to achieving this aim, providing stable transgenic expression in many different cell types and tissues.

1.2 Viral vectors

For the production of an efficacious and safe viral vector it is required at first to identify the crucial viral sequences for viral particle assembly, for viral genome package, and for *transgene* (see **glossary**) delivery to target cells. Then, dispensable genes are deleted from viral genome in order to reduce its pathogenicity and immunogenicity. At last, residual viral genome and transgene are integrated into the construct (Figure 2). Some viral vectors are able to integrate host genome unlike others that remain in an episomal form. Integrative vectors, like retroviruses and adeno-associated vectors are able to promote a persistent transgene expression. Otherwise, non-integrative vectors, like adenovirus whose viral DNA is maintained in an episomal form into infected cells, lead to a transient transgene expression. Each vector presents specific advantages and limitations that become more or less proper depending on the objective of its application (http://cmbi.bjmu.edu.cn/cmbidata/therapy/research/re02/021.htm; Osten et al., 2007; Howarth et al., 2010).

The ideal vector has not been described yet, but its characteristics should include:

- Easy and efficient production of high titers of viral particle;
- Absence of toxicity to target cells and undesirable effects like as immune response against the vector or the transgene;
- Capacity of site-specific integration, allowing a long-term transgene expression, as in cases of genetic disorders;
- Capacity of transduction of specific cell types;
- Infection of proliferative and quiescent cells.

Most vectors used for gene delivery are derived from human viral pathogens that have been made nonpathogenic by deleting essential viral genes. They usually have a broad tropism; therefore they can infect and deliver their encoded transgenes to a wide spectrum of cells and/or tissues.

Currently, the most efficient and commonly used viral vectors are adenovirus (Ad), adeno-associated virus (AAV), herpes simplex virus type 1-derived vectors (HSV-1), and retrovirus/lentivirus vectors (Table 1).

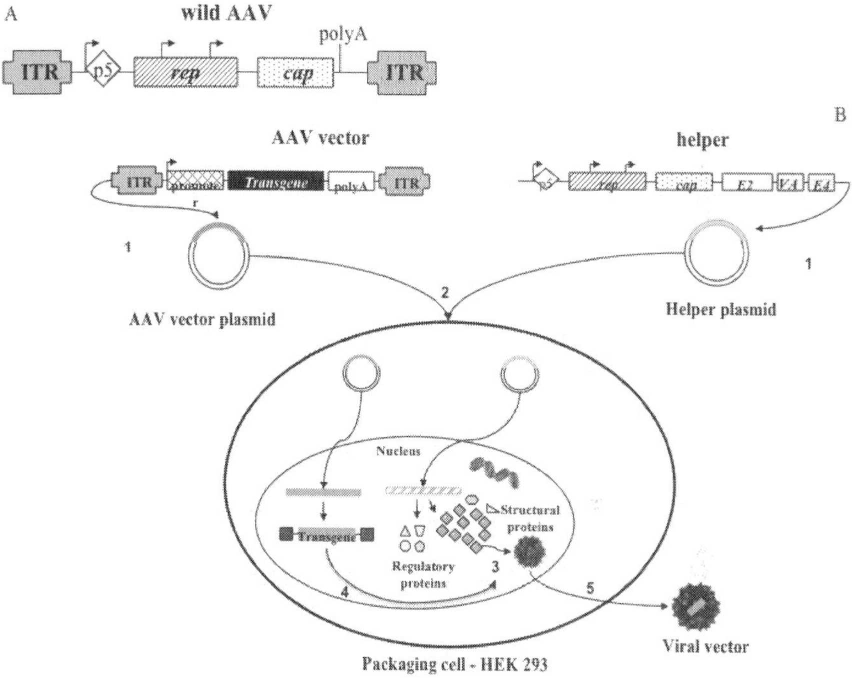

Fig. 2. Example of vector construction. (A) Wild AAV. (B) AAV recombinant vector production. AAV vector cassette, containing only the ITRs of the wild virus and the transgene and its promoter; helper cassette, containing the AAV rep and cap genes required for virus packaging and the Ad genes E2, VA and E4, required for virus replication. (1); co-transfection of both plasmids in HEK 293 packaging cells, whose genome contains the Ad E1 gene, which together with the other Ad required genes, supplied by the helper plasmid, allows the establishment of a productive infection (2); empty capsid is formed by AAV structural proteins assembly into the nucleolus (3); the replicated ssDNA viral genome with the transgene flanked by ITRs (vector plasmid) are packaged into the empty capsid in the nucleoplasm (4), giving rise to a non-replicative recombinant AAV vector virion (5). The regulatory proteins act especially in the replication and packaging processes (from Coura and Nardi, 2008).

Virus vector	Description	Advantages	Limitations	Applications
Adenovirus (Ad)	• Icosahedric • Non- enveloped • Genome capacity of 36 Kb • Entry into the cell by receptor binding • Episomal (non-integrative) • Capable of transducing postmitotic cells • Persistence of expression for weeks or months • Difficult production • Toxicity	• Easy propagation in high titers(10^{10}-10^{13}); • High transduction efficiency; • Infection of the majority of cell types; • Possibility of insertion of large DNA fragments (up to 8kb); • Infects both replicating and differentiated cells.	• High immunogenicity, inducing an important cellular and humoral immune response that can be fatal; • Transient expression • Viral proteins can be expressed in host following vector administration	Therapies that require transient gene expression: cancer therapy, angiogenesis induction and DNA vaccine production (due to its inflammatory and immunogenic properties);
Retrovirus (Retrovirus and Lentivirus)	• Globular • Enveloped • Genome capacity of 8 Kb • Entry into the cell by fusion • Integrative • Capable of transducing proliferative (retrovirus and lentivirus) and quiescent (lentivirus) cells • Persistence of expression for years • Easy production • Minimal toxicity	• Low immunogenicity; • Possibility of insertion of modest DNA fragments (up to 8 Kb); • High transduction efficiency; • Integrates into host genome resulting in sustained expression of vector • Extremely well studied system • Vector proteins not expressed in host	• Insertional mutagenesis (random integration) • Low titers (106- 107) • *In vivo* delivery remains poor. Effective only when infecting helper cell lines	• Genetic diseases of T cells and hematological diseases (Retrovirus); • HIV/AIDS • Optogenetics • Neuronal tracing
Adeno-associated virus (AAV)	• Icosahedric • Non- enveloped • Genome capacity of 4.7 Kb • Entry into the cell by receptor binding • Integrative • Capable of transducing postmitotic cells • Persistence of expression for years • Difficult production • No toxicity	• Low immunogenicity; • Easy propagation in high titers; • Infection of the majority of cell types; • Long-term gene expression	Limited capacity for DNA fragments insertion	• Genetic Diseases • Tumors • Neurological Diseases • Ocular Diseases • Cardiovascular Diseases • Others

Herpes Simplex Virus (HSV)	• Icosahedric • Enveloped • Genome capacity of 30 Kb • Entry into the cell by fusion • Episomal (non-integrative) • Capable of transducing postmitotic cells • Persistence of expression for weeks • Difficult production • Toxicity	• Large insert size • Could provide long- term CNS gene expression • High titer	• System currently under development • Current vectors provide transient expression • Low transduction efficiency • Toxicity	• Neuronal gene transfer, *in vitro* and *in vivo*.

Table 1. Main viral vectors systems and their characteristics.

1.2.1 Adenovirus

Adenoviruses (Ads) are medium-sized (90–100 nm), non-enveloped icosahedral viruses composed of a nucleocapsid and a double-stranded linear DNA genome. The viral genome is large, consisting of a single double-stranded DNA molecule 36 to 38 kilobases in size. Viral DNA replication and transcription are complexes, and viral replication and assembly occur only in the nucleus of infected cells. Mature virions are released by cellular disintegration.

There are 55 described serotypes in humans, which are responsible for 5–10% of upper respiratory infections in children, and many infections in adults as well.

Adenoviral infection is initiated by the virus binding to the cellular receptor. Internalization occurs via receptor-mediated endocytosis followed by release from the endosome. After endosomal release, the viral capsid undergoes disassembly while moving to the nuclear pore. Nuclear entry of the viral DNA is completed upon capsid dissociation, and the viral DNA does not integrate into the host genome but remains in an episomal state.

Since adenoviral replication depends on the E1A region of the viral genome, all recombinant adenoviral vectors have this region of its genome deleted, and are referred to as "replication-deficient" virus. Such vectors are capable of infecting a cell only once, no viral propagation occurs, and the infected cell does not die as a result of vector infection. As replication-deficient viruses are required, the 293 cell line is utilized for vector production. This is a human kidney cell line which has been stably transfected with the E1A region of the adenoviral genome. This allows the vector to be made and matured within the 293 cell, yet vectors prepared from this cell line will lack the E1A region and remain replication-deficient.

The non-integrative feature of adenovirus limits their use in basic research, although adenoviral vectors are occasionally used in *in vitro* experiments. Their primary applications are in gene therapy, especially oncolytic gene therapy, and vaccination (Ayuso et al., 2010).

1.2.2 Retrovirus

Retroviruses are globular enveloped virions ranging in diameter from 80 to 130 nm. The viral genome is encased within the capsid along with the proteins integrase and reverse transcriptase. The genome consists of two identical positive (sense) single-stranded RNA molecules ranging in size from 3.5 to 10 kilobases. Following cellular entry, the reverse transcriptase synthesizes viral DNA using the viral RNA as its template. The cellular machinery then synthesizes the complementary DNA which is then circularized and inserted into the host genome. Following insertion, the viral genome is transcribed and viral replication is completed. The majority of retroviruses are oncogenic although the degree to which they cause tumors varies from class to class.

As the inserted vector, called a provirus, remains in the genome, it passes on to the progeny of the cell when it divides. However, the site of integration is unpredictable, what can represent an obstacle in using of these vectors. The provirus can disturb the function of cellular genes and lead to activation of oncogenes promoting the development of cancer (insertional mutagenesis), which raises concerns for possible applications in gene therapy.

Retroviral vectors can either be replication-competent or replication-defective, even if replication-defective vectors are the most common choice. Replication-competent viral vectors contain all necessary genes for virion synthesis, and continue to propagate themselves once infection occurs. Because the viral genome for these vectors is much lengthier, the length of the actual inserted gene of interest is limited compared to the possible length of the insert for replication-defective vectors.

In addition to insertional mutagenesis, the primary drawback to use of retroviruses, such as the Moloney retrovirus, involves the requirement for cells to be actively dividing for transduction. As a result, cells such as neurons are very resistant to infection and transduction by retroviruses.

Lentiviruses are a subclass of Retroviruses. They have been adapted as gene delivery vehicles thanks to their ability to integrate into the genome of non-dividing cells, which is a unique feature of Lentiviruses as other Retroviruses can infect only dividing cells. Moreover, studies have shown that lentivirus vectors have a lower tendency to integrate in places that potentially cause cancer than other retroviral vectors.

For safety reasons lentiviral vectors never carry the genes required for their replication. To produce a lentivirus, several plasmids are transfected into a so-called packaging cell line, commonly HEK 293. One or more plasmids, generally referred to as packaging plasmids, encode the virion proteins, such as the capsid and the reverse transcriptase. Another plasmid contains the genetic material to be delivered by the vector. It is transcribed to produce the single-stranded RNA viral genome and is marked by the presence of the ψ (psi) sequence. This sequence is used to package the genome into the virion (Mátrai et al., 2010; Kumar & Woon-Khiong, 2011; Yi et al., 2011).

1.2.3 Adeno-associated virus

Adeno-associated viruses (AAV) belongs to the genus Dependovirus and family Parvoviridae. It is currently not known to cause disease. The Parvoviridae family is characterized by small, ichosaedral and non-enveloped virus whose genome is a single stranded DNA. AAV is one of the smallest viruses with a capsid of approximately 22 nm and one of the most spread of this family, leading to seropositivity in more than 80% of human population (serotype 2). Despite this high seroprevalence, the virus has not been linked to any human illness, causing a very mild immune response. Because a co-infecting helper virus is usually required for a productive infection to occur, AAV serotypes are ascribed to a separate genus in the Parvoviridae family designated Dependovirus.

The wild AAV has a linear single-stranded DNA genome of approximately 4.7 Kb of either plus (sense) or minus (anti-sense) polarity. The AAV2 DNA termini consist of a 145 nucleotide-long inverted terminal repeat (ITR) that forms a characteristic T-shaped hairpin structure. These ITRs are important in the site-specific integration of AAV DNA into a specific site in chromosome 19. The ability of wild-type AAV to selectively integrate into chromosome 19 made them an attractive candidate for the production of a gene therapy vector that could do the same. The ITRs act as DNA replication origin, as well as signal for package and integration. In addition, they also act as regulator element for wild AAV gene expression. The ITRs flank the two viral genes rep (replication) and cap (capsid) encoding nonstructural and structural proteins, respectively. The virus does not encode a polymerase relying instead on cellular polymerase activity to replicate its DNA (Ni et al., 1998).

Adeno-associated vectors are prepared by replacing the capsid genes with the gene of interest. Construction of AAV vectors consists of the recombinant AAV vector plasmid DNA, and a non-rescuable AAV helper plasmid, which encodes for the AAV capsid proteins. Also required is either wild-type adenovirus or HSV and cell line for viral propagation. Unlike the previous vector systems described, the cell line need not contain any portion of the AAV genome since all required AAV genome elements could be provided by the two plasmids. Cells are first infected with the wild-type adenovirus or HSV, and then both the recombinant AAV vector plasmid DNA and the non-rescuable AAV helper plasmid are co-transfected into the cells. The cells produce mature recombinant AAV vectors as well as wild-type adenovirus or HSV. The wild-type adenovirus or HSV is removed by either density gradient centrifugation or heat inactivation.

These vectors have been designed to produce a gene therapy vector with site-specific integration and the ability to infect multiple cell types. Unfortunately, this has not been the case to date for these vectors. Current research focuses on how to regain the site-specific integration sequences into the recombinant vector. These vectors do offer some advantages over other vector systems which include the lack of initiating an immune response, their stability and ability to infect a variety of dividing and non-dividing cells. Unfortunately, they cannot incorporate genes larger than 5 kb and must be closely screened for adenoviral or HSV contamination (Berns, 1996; Grimm & Kay, 2003; Coura & Nardi, 2007; Coura & Nardi, 2008; Mezzina & Merten, 2011).

1.2.4 Herpes simplex virus

Herpes simplex viruses are members of the herpes virus family, Herpesviridae, that infect humans. They are ubiquitous neurotropic and neuroinvasive viruses that persist in the body by becoming latent and hiding from the immune system in the cell bodies of nerves. Latency is defined as a state in which viral DNA is maintained within the cell nucleus in the absence of any viral replication. The structure of herpes viruses consists of a relatively large double-stranded, linear DNA genome encased within an icosahedral capsid, which is wrapped in a lipid bilayer envelope. The HSV genome has been entirely sequenced and is rather extensively studied. As a result, currently, a general knowledge exists of which genes and DNA sequences may be deleted and at which sites foreign DNA may be inserted into the viral genome. These studies also have defined the minimal requirements for viral replication and packaging.

Vectors derived from Herpes simplex virus (HSV) have some unique features. The vectors have a wide host range and cell tropism, infecting almost every cell type in most vertebrates. In addition, the natural property of the virus to infect and establish latent infection indefinitely in post-mitotic neurons has generated substantial interest in using it to deliver therapeutic genes to the nervous system.

The two main strategies for HSV-based vectors in use today are genetically-engineered viruses and plasmid derived "amplicon" vectors. The first strategy involves the construction of recombinant viruses containing deletions in one or more viral genes whose expression is essential for viral replication, resulting in replication-incompetent vectors.

The second strategy involves the use of plasmid-derived vectors containing HSV-1 origins of DNA replication and DNA packaging signals that enable multiple copies of the vector genomes to be packaged into helper virus virions. Helper viruses can be either recombinant viruses containing a deletion within an essential viral gene or viruses containing temperature-sensitive mutations that prevent replication at 37° C (normal body temperature). In the case of the former, the replication of the helper virus and packaging of the amplicon vector DNA must occur in a cell line capable of complementing the mutations in the helper virus.

Regardless of the vector system used, two primary goals must be achieved to enable long-term gene expression in neuronal cells. The first goal involves the construct of mutant vectors which themselves are noncytotoxic to cells. Several studies have noted active expression of a foreign gene by HSV vector constructs, which subsequently became inactivated. Reasons for this are not completely apparent, but evidence suggests that the inactivation is a result of cytotoxic effects induced by vector systems.

The second goal involves designing stable, active promoters capable of expressing appropriate levels of the foreign protein. The specific promoter involved in individual therapies may change according to the type, status and activity of the neuronal cell of interest.

Currently, many studies have demonstrated long-term stable transgene expression in the nervous system. In addition, preclinical studies on models of neurological disease, such as glioma, peripheral neuropathy, chronic pain and neurodegeneration, show encouraging results (Jenkins & Turner, 1996; de Silva & Bowers, 2009; Glorioso & Fink, 2009; Manservigi et al., 2010; Fraefel et al., 2011; Goins et al., 2011).

1.3 Viral vectors in neurobiology

Gene transfer into the central nervous system (CNS) shows great promise for basic and clinical research in neurosciences. As the brain presents a high level of structural complexity, it is a complicated target to be accessed and for genetic manipulation.

Currently, viruses are the most widely used vehicles for gene transfer into the adult mammalian brain. However, there does not exist a "universal ideal vector" and each basic or clinical approach may require a specific set of technical hurdles to overcome. Several viral vectors have been studied. Each one has shown great advantages and disadvantages depending basically on the subset of target cells and the specificities of each research or clinical indication.

The number of suitable vectors for basic research surpasses those being used in clinical trials. Currently, the most widely used vectors in neurobiology are AAV, HSV and lentiviral vectors. AAV presents several serotypes with specific cell tropism. AAV serotype 2 (AAV2), for example, infects neurons preferentially, but seems to not infect all types of neurons equally well. Other serotypes, as AAV4 and AAV5, show distinct tropism and diffusion properties. The construction of vectors combining more than one serotype and using specific cell promoters, allows genetic manipulation of specific sets of neurons, with more sustained and effective transgene delivery and expression. AAV vectors are highly effective for gene delivery and are non-toxic. The main limitation of these vectors is its relatively small gene capacity (McCown, 2011).

HSV is neurotrophic and shows a highly efficient retrograde and anterograde transport within the CNS, being able of entering in a benign latent state. HSV vectors have a large transgene capacity and can assure long-lasting transgene expression. However, the main disadvantage of this type of vector is its cell toxicity and low transduction efficiency. Currently, other variants that try to surpass these limitations have been developed.

Lentivirus vectors have a modest packaging capacity, induce minimal immunological response and can produce long-term transgene expression. In addition, envelope-engineered vectors can show broad cell tropism. On the other hand, these vectors show poor *in vivo* delivery and present the risk of insertional mutagenesis. However, as lentiviruses mostly transduce terminally differentiated cells, the risk of insertional mutagenesis is less important than observed for other retroviruses (Kaplitt & Pfaff, 1996; Davidson & Breakefield, 2003; Howarth et al., 2010; Snyder et al., 2010).

2. Therapeutic and research applications

Viral vectors were originally developed as an alternative to *transfection* (see **glossary**) of naked DNA for molecular genetics experiments. Compared to traditional methods such as calcium phosphate precipitation, *transduction* (see **glossary**) can ensure that nearly 100% of cells are infected without severely affecting cell viability. Furthermore, some viruses integrate into the cell genome facilitating stable expression.

Protein coding genes can be expressed using viral vectors, commonly to:

1. Increase concentration of a certain protein and study its function (over-expression studies);
2. Antagonize function of a certain protein (expression of dominant negative proteins and RNAi constructs);
3. Make the cell produce fluorescent indicator proteins (for example, EGFP or Ca2+ sensitive proteins). These may be used to monitor various variables within the living cells (tracing and *in vivo* imaging);
4. Control neuronal excitability using light-sensitive ion channels (optogenetics);
5. Pre-clinical and clinical gene therapy approaches.

Therefore, viral vectors constitute an important tool and present extraordinary opportunities for basic and clinical research, particularly in neurosciences.

2.1 Gene expression *in vitro* and *in vivo*

Viral vectors have emerged as an important tool for manipulating gene expression in the adult mammalian brain. The adult brain is composed largely of non-dividing cells, and therefore DNA viruses have become the vehicle of choice for neurobiologists interested in somatic gene transfer. By re-expressing an absent protein or by overexpressing a certain protein (i.e increasing its concentration), one can unravel its functions and possible process that it underlies (Howarth et al., 2010).

2.2 Tracing

Detailed knowledge of the complex anatomical interconnections of the central nervous system (CNS) plays an important role in the understanding of brain function, both in physiological and in pathological conditions. For this, *in vivo* tracing of neural tracts constitutes an important tool. Some viral vectors systems, especially lentivirus, are largely used to this end. The idea is to construct vectors expressing a marker protein, like GFP that are anterogradely (from the cell soma to the *neurites* - see **glossary**) or retrogradely (from the neurites to the cell soma) transported along the neurons (Figure 3). In this manner, one can inject an anterograde vector into a certain brain region in order to see where its projections are issued. Similarly, using a retrograde vector, one can identify from where come its inputs (Figure 4) (Masamizu et al., 2011).

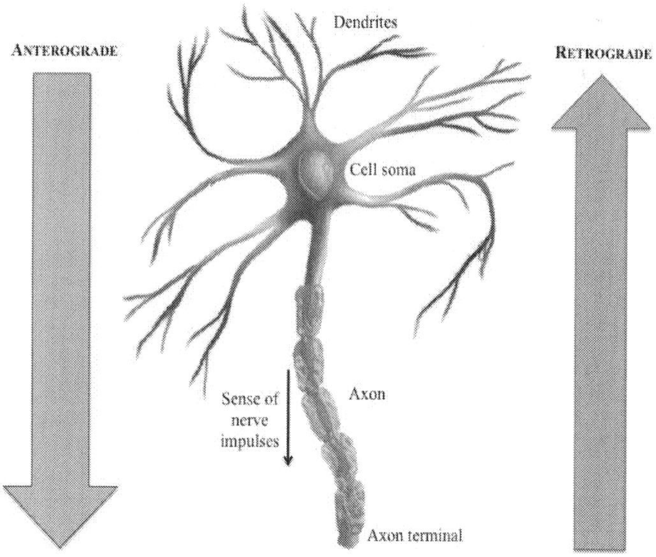

Fig. 3. Transport anterograde and retrograde. The former is from de cell soma to the axon terminals. This is also the sense of the nerve impulsions. Conversely, retrograde transport is from the axon terminals to the cell soma. Then, if one injects an anterograde vector in a certain brain region, we will observe to where it issues its projections. On the other hand, injection of a retrograde vector, allows us to know from where come the inputs of the injected brain region.

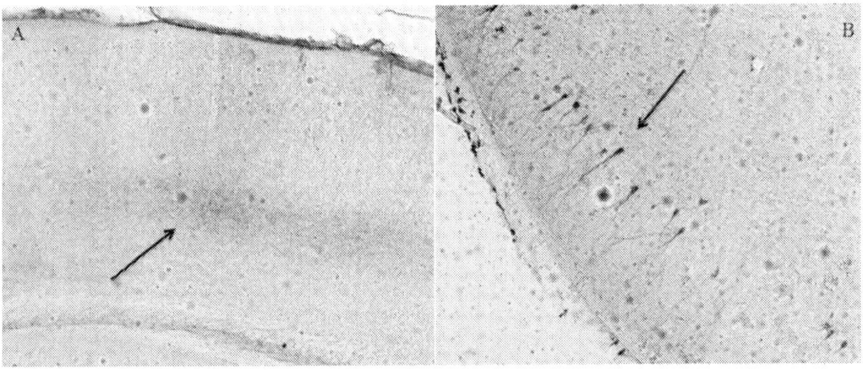

Fig. 4. (A) Optical microscopy of an anterograde vector, showing axon projections from the prefrontal cortex. (B) Optical microscopy of a retrograde vector, showing cell soma that inputs into the prefrontal cortex.

2.3 Optogenetics

Optogenetics is a new and promising gene and neuroengineering tool, which allows the control of the activity of defined populations of neurons. With this technology, by the combination of genetic and optical methods, imaging or control of specific events in target cells of living tissues with temporal precision became possible. The optical part consists in the use of specific light-sensitive proteins expressed in the neurons of interest. These are channel proteins that have either been rendered responsive to light or that are inherently light-sensitive, making possible the manipulation, that is, activation or inhibition, of neuronal activity.

The genetic methods concern to the construction of *expression cassettes*, using cell type specific *promoters* (see **glossary**). Specifically coupling the DNA sequence of the light-sensitive protein with a cell-specific promoter, one can drive protein expression in neurons defined by a common genetic identity and, possibly, common functional roles.

Optogenetics has provided a tool that may overcome some of the limitations of traditional neuromodulation techniques. Activation or inhibition of specific neuronal populations with different wavelengths of light opens up possibilities for modulating neural circuits with previously unimagined levels of precision.

Optogenetics have advanced rapidly since 2005, when it has been demonstrated that neurons expressing the ChR2 (channelrhodopsin-2) protein can generate action potentials when illuminated by light with a specific wavelength (Boyden et al. 2005). This protein is derived from a unicellular alga that was engineered for stable membrane expression into mammalian brain. This is a light-sensitive cation channel that triggers *depolarizing action potentials* (see **glossary**) when exposed to 470 nm blue light. Currently, ChR2 protein variants have been generated in order to improve its functionality, especially concerning deactivation kinetics and activation lasting.

Another light-sensitive protein largely used as an optogenetics tool is a bacteria-derived chloride pump, called halorhodopsin (NpHR). This protein provokes neuronal activity inhibition, by generating a fast chloride ion influx whereby it leads to membrane *hyperpolarization* (see **glossary**), when activated by 570 nm yellow light. As for ChR2, more efficient variants of NpHR have been generated.

As ChR2 and NpHR proteins are activated by separate wavelengths of light, activation or inhibition of action potential activity in ChR2- and NpHR-expressing cells can be independently controlled. Moreover, co-expression of ChR2 and NpHR within a single neuron allows bidirectional control of membrane action potential.

These light-sensitive proteins, both belonging to the family of opsin genes, can be genetically targeted into specific brain cell types using stereotactically injected viral vectors. Lentiviral and adeno-associated viral (AAV) vectors are both suitable for such gene transfer. Notably, AAV has been considered a safer vector for CNS gene transfer and has been used in several clinical trials, because of its innocuous and non-pathogenic features.

Other expression systems were developed that are more suitable for basic and preclinical researches than for clinical trials. This includes generation of transgenic mice expressing ChR2 in a subset of neurons. Another approach consists in the use of Cre-recombinase *knock-in mice* (see **glossary**) injected with Cre-activated AAV vectors encoding opsin genes (Henderson et al., 2009; Fiala et al., 2010; Knöpfel et al., 2010; Mancuso et al., 2010; Kravitz & Kreitzer, 2011; LaLumière, 2011; Yizhar et al., 2011).

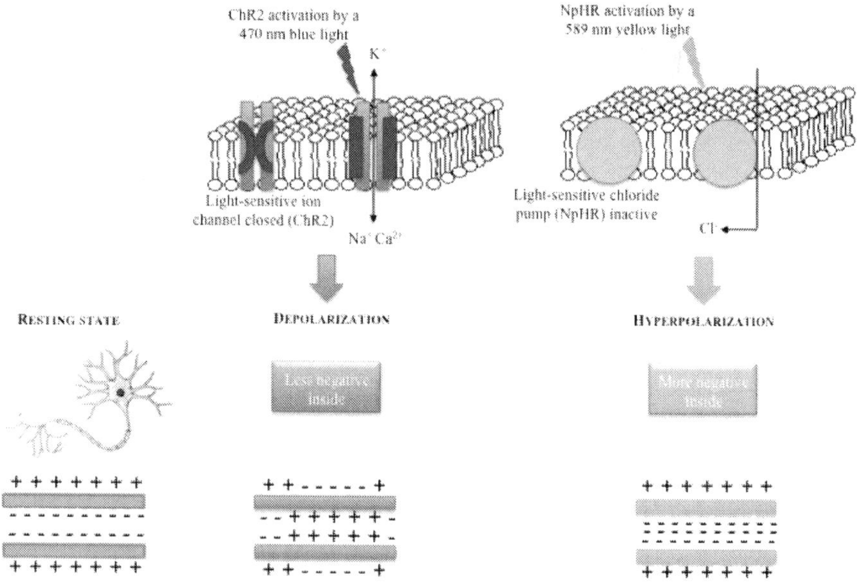

Fig. 5. Sensitive-light proteins and the basis of optogenetics technology.

These techniques have been used in the analysis of neural circuits, activation of reflexive behavior, induction of behavior plasticity, linking cell activity to behavior, among others. As an example, using a combination of Cre-inducible ChR2 with parvalbumin-positive interneuron-specific or pyramidal cell-specific Cre driver cell lines, it was identified that parvalbumin fast-spiking interneurons have a specific and important role in the generation of gamma-oscillations in the barrel cortex and that perceptual decisions and learning can be controlled by a subset of excitatory neurons, in mice (Cardin et al. 2009; Sohal et al., 2009).

Currently, the possible application of optogenetic tools with therapeutic approaches for neurological conditions just starts to be considered and investigated. Possible targets would be activation and recovery of breathing, suppression of seizure-like activity, amelioration of parkinsonian symptoms, recovery of blindness and optical deep brain stimulation. Yet, peripheral nervous system targets could also be plausible (Nagel et al., 2005; Gradinaru et al., 2009; Adamantidis et al., 2010; Dani et al., 2010; Gradinaru et al., 2010; Zhang et al., 2010; Depuy et al., 2011; Kokaia & Sørensen, 2011; Peled, 2011; Tønnesen et al., 2011).

2.4 Gene therapy

Gene therapy (GT) constitutes a therapeutic intervention based on modifications of the genetic material of target cells, by either correcting genetic defects or overexpressing therapeutically useful proteins. Initially designed to definitely correct monogenic disorders such as cystic fibrosis, severe combined immunodeficiency or muscular dystrophy, gene therapy has evolved into a promising therapeutic modality for a diverse array of diseases. Targets are expanding and currently include not only genetic, but also many acquired diseases, such as cancer, tissue degeneration or infectious diseases. Over the past years, significant progress was made concerning to *enabling technologies* (see **glossary**), molecular understanding of several of these diseases and manufacturing of vectors. The basic prerequisites to GT success include therapeutically suitable genes, appropriate gene delivery systems and proof of safety and efficacy. Among them, gene delivery systems constitute a crucial requirement. Depending on the duration planned for the treatment, type and location of target cells, and whether they undergo division or are quiescent, different vectors may be used, involving non-viral methods, non-integrating viral vectors or integrating viral vectors. Advances, especially, in viral vectors engineering and improved gene regulatory systems to facilitate and tightly control therapeutic gene expression have leading to GT progress (Coura & Nardi, 2008).

The first gene therapy clinical trial was carried out in 1989, in patients with advanced melanoma, using tumor-infiltrating lymphocytes modified by retroviral transduction. In the early 90's, a clinical trial with children with severe combined immunodeficiency (SCID) was also performed, by retrovirus transfer of the deaminase adenosine gene to lymphocytes isolated from these patients. Since then, more than 5,000 patients have been treated in more than 1,000 clinical protocols all over the world (http://www.wiley.co.uk/genetherapy/clinical). Despite the initial enthusiasm, however, the efficacy of gene therapy in clinical trials has not been as high as expected; a situation further complicated by ethical and safety concerns (Coura & Nardi, 2008). Further studies are being developed to solve these limitations.

2.4.1 Brain gene therapy

The advances of modern molecular biology and *in vivo* gene therapy have challenged neuroscientists with the potential prospect of genetically manipulating post-mitotic neurons. The ability to alter gene expression in these cells would open the door towards potential therapies for several disorders such as Parkinson's disease, Huntington's disease and amyotrophic lateral sclerosis. Gene therapy using viral-based vectors has received considerable attention and represents a major focus of ongoing research. Viral vectors using several different human viruses such as adeno-associated viruses, herpes viruses and lentiviruses are currently being developed for brain gene therapy purposes. Gene therapy directed towards neuronal cells however, presents unique problems. These problems include the genetic manipulation of post-mitotic (*i.e.*, non-dividing) cells, the ability to specifically infect neurons, long-term maintenance of the vector DNA and expression of the target gene within the neuronal cells. Herpesviruses, particularly herpes simplex virus type 1, have unique characteristics of infection, replication and pathogenesis which make them potentially ideal candidates for the development of viral vectors capable of altering endogenous gene expression or delivery of foreign genes both *in vivo* and *in vitro*.

In the same way, AAV vectors present great potential and appear to be much more promising for a wide range of gene therapy approaches. Current studies have been leading to great improvements in AAV vector and expression cassette design and novel AAV serotypes have been identified, that have improved AAV vectors efficacy (Kaspar et al., 2002; Coura & Nardi, 2007; McCown, 2011).

Vectors based on minimal self-inactivating (SIN) and pseudotyping lentiviruses have been considered as relevant vector for research and clinical applications. In these vectors, the entire coding regions of the virus are removed and provided in *trans* from separated expression cassettes, so that they present less than 1 kb of viral genome and express only the transgene cassette. In addition, the use of an envelope from another virus (pseudotyping), like the vesicular stomatitis virus (VSV-G), can give vectors broad species and tissue tropism. Moreover, innate and adaptive immune responses against these lentiviral vectors are not significant. All together, these features are encouraging their use for CNS applications (Escors & Breckpot, 2010).

All these vectors have been largely used in pre-clinical and/or clinical trials. In many indications, both AAV and lentiviral vectors are being assessed. This is the case for gene therapy approaches in epilepsy, multiple sclerosis, Alzheimer disease, diabetes, Parkinson's disease, chronic pain, lysosomal storage disorders, amyotrophic lateral sclerosis, brain ischemia, seizure, Huntington disease, and others. HSV vectors have also been used for some of these indications, but currently, they have been frequently applied for gene therapy approaches in brain tumors, as is the case for adenovirus vectors as well (Maingay et al., 2005; Wong et al., 2006; Robinson et al., 2007; Lowery et al., 2009; Shih et al., 2009; Björklund et al., 2010; Manfredsson & Mandel, 2010; Eskenazi & Neumaier, 2011; Jacobs & Wang, 2011; Van der Perren et al., 2011; Vande Velde et al., 2011; Thaci et al., 2011).

3. Summary

Viruses are intracellular parasites with simple DNA or RNA genomes with a unique capacity to enter the cell and to replicate their genome. This has given rise to the possibility to construct vectors containing the viral envelop and a recombinant genome, so that these vectors could be able to deliver genetic material into cells. Viral vectors hold promise for basic and clinical research in neurosciences and currently the vehicles of choice for gene transfer into the adult mammalian brain. Main viral vectors exploited for use in neurobiology are adeno-associated vectors, herpes simplex vectors and lentiviral vectors. Each has both advantages and disadvantages depending basically on the subset of target cells and the specificities of each research or clinical indication.

The spectrum of research and therapeutic applications of viral vectors increase to the extent that advances are made in this area. Currently, these applications include functional studies by gene expression *in vitro* and *in vivo*; neuronal tracing, allowing unraveling anatomical neuronal interconnections; neuromodulation, using optogenetic tools and gene therapy.

Nevertheless, for each approach there are still a lot of technical hurdles to overcome. However, advancement in this field will most likely lead to new tools being created with the ability to overcome current limitations.

4. Glossary

Term	Definition
Gene expression	The process of formation of messenger RNA (mRNA) of a DNA template which then is translated into the sequence of aminoacids at the ribosome to make proteins.
Transgene	A foreign gene introduced into the cell (for example by a viral vector).
Transfection	Process of delivering a foreign gene into the target cell, by a non-viral vector or naked DNA.
Transduction	Process of delivering a foreign gene into the target cell, by a viral vector.
Expression cassette	A piece of DNA containing elements (promoter, coding part and polyadenylation signal) necessary for expression of a transgene.
Promoter	Region of DNA that facilitates the transcription of a particular gene. Promoters are located near the genes they regulate, on the same strand and typically upstream (towards the 5′ region of the sense strand).
Cre	Cre is a 38 kDa recombinase protein from bacteriophage P1 which mediates intramolecular (excisive or inversional) and intermolecular (integrative) site specific recombination between loxP sites. A loxP consists of two 13 bp inverted repeats separated by an 8 bp asymmetric spacer region. The Cre's DNA excising capability can be used to turn on a foreign gene by cutting out an intervening stop sequence between the promoter and the coding region of the transgene.
Gene knockout	It is a genetic technique in which one of an organism's gene is made inoperative, basically by gene deletion or disruption.
Knock-in	Genetic engineering method that involves the insertion of a protein coding sequence at a particular locus in an organism's chromosome.
Neurites	Refers to any projection from the cell body of a neuron. This projection can be either an axon or a dendrite.
Depolarization	Change in a cell's membrane potential, making it more positive or less negative. A large enough depolarization may result in an action potential that triggers excitation of the neuron activity.
Hyperpolarization	Change in a cell's membrane potential, making it more negative, inhibiting the rise of an action potential and, consequently inhibiting neuronal activity.
Enabling technology	Set of new processes and new techniques that enable the development and improvement of existing technologies.

5. References

Abdellatif, A.A., Pelt, J.L., Benton, R.L., Howard, R.M., Tsoulfas, P., Ping, P., Xu, X.M. & Whittemore, S.R. (2006) Gene delivery to the spinal cord: comparison between lentiviral, adenoviral, and retroviral vector delivery systems. *J Neurosci Res.*, Vol. 84, No. 3, pp. 553-567.

Adamantidis, A., Carter, M.C. & de Lecea, L. (2010) Optogenetic deconstruction of sleep-wake circuitry in the brain. *Front Mol Neurosci.*, Vol. 2, pp. 31.

Ayuso, E. Mingozzi, F. & Bosch, F. (2010) Production, purification and characterization of adeno-associated vectors. Curr *Gene Ther.*, Vol. 10, No. 6, pp. 423-436.

Berns, K. I. (1996) Parvoviridae: The virus and their replication. In: Fields, B.N., Knipe, D.M. & Howley, P.M. (eds). Fields in Virology. Lippincott - Raven, Philadelphia, pp.2173–2197.

Björklund, T., Cederfjäll, E.A. & Kirik, D. (2010) Gene therapy for dopamine replacement. *Prog Brain Res.*, Vol. 184, pp. 221-235.

Cardin, J.A., Carlén, M., Meletis, K., Knoblich, U., Zhang, F., Deisseroth, K., Tsai, L.H. & Moore, C.I. (2009) Driving fast-spiking cells induces gamma rhythm and controls sensory responses. *Nature*, Vol. 459, No. 7247, pp. 663-667.

Coura, R.S. & Nardi, N.B. (2007) The state of the art of adeno-associated virus-based vectors in gene therapy. *Virology Journal*, vol. 16, 4, pp.99.

Coura, R.S. & Nardi, N.B. (2008) A role for adeno-associated viral vectors in gene therapy. *Genet. Mol. Biol.*, Vol.31, No.1, pp. 1-11.

Dani, A., Huang, B., Bergan, J., Dulac, C. & Zhuang, X. (2010) Superresolution imaging of chemical synapses in the brain. *Neuron*, Vol. 68, No. 5, pp. 843-856.

Davidson, B.L. & Breakefield, X.O. (2003) Viral vectors for gene delivery to the nervous system. *Nat Rev Neurosci*, Vol. 4, No. 5, pp. 353-364.

de Silva, S. & Bowers, W.J.(2009) Herpes Virus Amplicon Vectors. *Viruses*, Vol. 1, No. 3, pp. 594-629.

Depuy, S.D., Kanbar, R., Coates, M.B., Stornetta, R.L. & Guyenet, P.G. (2011) Control of breathing by raphe obscurus serotonergic neurons in mice. *J Neurosci.*, Vol.31, No. 6, pp.1981-1990.

Escors D. & Breckpot, K. (2010) Lentiviral vectors in gene therapy: their current status and future potential. *Arch Immunol Ther Exp (Warsz)*, Vol. 58, No. 2, pp. 107-119.

Eskenazi, D. & Neumaier, J.F. (2011) Increased expression of 5-HT(6) receptors in dorsolateral striatum decreases habitual lever pressing, but does not affect learning acquisition of simple operant tasks in rats. *Eur J Neurosci.*, Vol.34, No. 2, pp. 343-351.

Fiala, A., Suska, A. & Schlüter, O.M. (2010) Optogenetic approaches in neuroscience. *Curr Biol.*, Vol. 20, No. 20, pp. R897-903.

Fraefel, C., Marconi, P. & Epstein, AL. (2011) Herpes simplex virus type 1-derived recombinant and amplicon vectors. *Methods Mol Biol.*, Vol. 737, pp. 303-343.

Glorioso, J.C. & Fink, D.J. (2009) Herpes vector-mediated gene transfer in the treatment of chronic pain. *Mol Ther.*, Vol. 17, No. 1, pp. 13-18.

Goins, W.F., Krisky, D.M., Wechuck, J.B., Wolfe, D., Huang, S. & Glorioso, J.C. (2011) Generation of replication-competent and -defective HSV vectors. *Cold Spring Harb Proto.*, in press.

Gradinaru, V., Mogri, M., Thompson, K.R., Henderson, J.M. & Deisseroth, K. (2009) Optical deconstruction of parkinsonian neural circuitry. *Science*, Vol. 324, No. 5925, pp. 354-359.

Gradinaru, V., Zhang, F., Ramakrishnan, C., Mattis, J., Prakash, R., Diester, I., Goshen, I., Thompson, K.R. & Deisseroth, K. (2010) Molecular and cellular approaches for diversifying and extending optogenetics. *Cell*, Vol.141, No. 1, pp.154-165.

Grimm, D. & Kay, M. A. (2003) From virus evolution to vector revolution: use of naturally occurring serotypes of adeno-associated virus (AAV) as novel vectors for human gene therapy. *Curr. Gene Ther.*, Vol. 3, pp. 281–304.

Henderson, J.M., Federici, T. & Boulis, N. (2009) Optogenetic neuromodulation. *Neurosurgery*, Vol. 64, No. 5, pp.796-804.

Howarth, J.L., Lee, Y.B. & Uney, J.B. (2010) Using viral vectors as gene transfer tools (Cell Biology and Toxicology Special Issue: ETCS-UK 1 day meeting on genetic manipulation of cells). *Cell Biol Toxicol*, Vol. 26, pp. 1-20.

http://cmbi.bjmu.edu.cn/cmbidata/therapy/research/re02/021.htm

http://www.wiley.co.uk/genetherapy/ clinical

Jacobs, F. & Wang, L. (2011) Adeno-Associated Viral Vectors for Correction of Inborn Errors of Metabolism: Progressing Towards Clinical Application. *Curr Pharm Des.*, in press

Jenkins, F.J. & Turner, S.L. (1996) Herpes simplex virus: a tool for neuroscientistis. *Frontiers in Bioscience*, Vol. 1, pp. d241-247.

Kaplitt, M.G. & Pfaff, D.W. (1996) Viral Vectors for Gene Delivery and Expression in the CNS. *Methods,* Vol. 10, No. 3, pp. 343-350.

Kaspar, B.K., Vissel, B., Bengoechea, T., Crone, S., Randolph-Moore, L., Muller, R., Brandon, E.P., Schaffer, D., Verma, I.M., Lee, K.F., Heinemann, S.F. & Gage, F.H. (2002) Adeno-associated virus effectively mediates conditional gene modification in the brain. *Proc Natl Acad Sci U S A*, Vol. 99, No. 4, pp. 2320-2325.

Knöpfel, T., Lin, M.Z., Levskaya, A., Tian, L., Lin, J.Y. & Boyden, E.S. (2010) Toward the second generation of optogenetic tools. *J Neurosci.*, Vol. 30, No. 45, pp. 14998-15004.

Kokaia, M. & Sørensen, A.T. (2011) The treatment of neurological diseases under a new light: the importance of optogenetics. *Drugs Today (Barc)*, Vol. 47, No. 1, pp. 53-62.

Kravitz, A.V. & Kreitzer, A.C. (2011) Optogenetic manipulation of neural circuitry in vivo. *Curr Opin Neurobiol.*, Vol. 21, No. 3, pp. 433-439.

Kumar, P., Woon-Khiong, C. (2011) Optimization of lentiviral vectors generation for biomedical and clinical research purposes: contemporary trends in technology development and applications. *Curr Gene Ther.*, Vol.11, No. 2, pp. 144-153.

LaLumiere, R.T. (2011) A new technique for controlling the brain: optogenetics and its potential for use in research and the clinic. *Brain Stimul.*, Vol. 4, No. 1, pp. 1-6.

Lowery, R.L., Zhang, Y., Kelly, E.A., Lamantia, C.E., Harvey, B.K. & Majewska, A.K. (2009) Rapid, long-term labeling of cells in the developing and adult rodent visual cortex using double-stranded adeno-associated viral vectors. *Dev Neurobiol.*, Vol. 69, No. 10, pp. 674-688.

Maingay, M., Romero-Ramos, M. & Kirik, D. (2005) Viral vector mediated overexpression of human alpha-synuclein in the nigrostriatal dopaminergic neurons: a new model for Parkinson's disease. *CNS Spectr.*, Vol. 10, No. 3, pp. 235-244.

Mancuso, J.J., Kim, J., Lee, S., Tsuda, S., Chow, N.B. & Augustine, G.J. (2011) Optogenetic probing of functional brain circuitry. *Exp Physiol.*, Vol. 96, No. 1, pp. 26-33.

Manfredsson, F.P. & Mandel, R.J. (2010) Development cf gene therapy for neurological disorders. *Discov Med.*, Vol. 9, No. 46, pp. 204-211.

Manservigi, R., Argnani, R. & Marconi, P. (2010) HSV recombinant vectors for gene therapy. *The Open Virology Journal*, Vol. 4, pp. 123-156.

Masamizu ,Y., Okada, T., Kawasaki, K., Ishibashi, H., Yuasa, S., Takeda, S., Hasegawa, I. & Nakahara, K. (2011) Local and retrograde gene transfer into primate neuronal pathways via adeno-associated virus serotype 8 and 9. *Neuroscience*, in press.

Mátrai, J., Chuah, M.K. & VandenDriessche, T. (2010) Recent advances in lentiviral vector development and applications. Mol Ther., Vol. 18, No. 3, pp. 477-490.

McCown, T.J. (2011) Adeno-Associated Virus (AAV) Vectors in the CNS. *Curr Gene Ther.*, Vol. 11, No. 3, pp. 181-188.

Mezzina, M. & Merten, O.W. (2011) Adeno-associated viruses. *Methods Mol Biol.*, Vol. 737, pp.211-234.

Nagel, G., Brauner, M., Liewald, J.F., Adeishvili, N., Bamberg, E. & Gottschalk, A. (2005) Light activation of channelrhodopsin-2 in excitable cells of Caenorhabditis elegans triggers rapid behavioral responses. *Curr Biol.*, Vol. 15, No. 24, pp. 2279-2284.

Osten, P., Grinevich, V. & Cetin, A. (2007) Viral vectors: a wide range of choices and high levels of service. *Handb Exp Pharmacol.*, Vol. 178, pp. 177-202.

Peled, A. (2011) Optogenetic neuronal control in schizophrenia. *Med Hypotheses*, Vol. 76, No. 6, pp. 914-921.

Petruska, J.C., Kitay, B., Boyce, V.S., Kaspar, B.K., Pearse, D.D., Gage, F.H. & Mendell, L.M. (2010) Intramuscular AAV delivery of NT-3 alters synaptic transmission to motoneurons in adult rats. *Eur J Neurosci.*, Vol. 32, No. 6, pp. 997-1005.

Robinson, S., Rainwater, A.J., Hnasko, T.S. & Palmiter, R.D. (2007) Viral restoration of dopamine signaling to the dorsal striatum restores instrumental conditioning to dopamine-deficient mice. *Psychopharmacology (Berl)*, Vol. 191, No. 3, pp. 567-578.

Shih, C.S., Laurie, N., Holzmacher, J., Spence, Y., Nathwani, A.C., Davidoff, A.M. & Dyer, M.A. (2009) AAV-mediated local delivery of interferon-beta for the treatment of retinoblastoma in preclinical models. *Neuromolecular Med.*, Vol. 11, No. 1, pp. 43-52.

Snyder, B.R., Boulis, N.M. & Federici, T. (2010) Viral vector-mediated gene transfer for CNS disease. *Expert Opin Biol Ther.*, Vol. 10, No. 3, pp. 381-394.

Sohal, V.S., Zhang, F., Yizhar, O. & Deisseroth, K. (2009) Parvalbumin neurons and gamma rhythms enhance cortical circuit performance. *Nature*, Vol. 459, No. 7247, pp. 698-702.

Tannemaat, M.R., Boer, G.J., Eggers, R., Malessy, M.J. & Verhaagen, J. (2009) From microsurgery to nanosurgery: how viral vectors may help repair the peripheral nerve. *Prog Brain Res.*, Vol. 175, pp. 173-186.

Thaci, B., Ulasov, I.V., Wainwright, D.A. & Lesniak, M.S. (2011) The challenge for gene therapy: innate immune response to adenoviruses. *Oncotarget.*, Vol. 2, No. 3, pp. 113-121.

Tønnesen, J., Parish, C.L., Sørensen, A.T., Andersson, A., Lundberg, C., Deisseroth, K., Arenas, E., Lindvall, O. & Kokaia, M. (2011) Functional integration of grafted neural stem cell-derived dopaminergic neurons monitored by optogenetics in an in vitro Parkinson model. *PLoS One*, Vol. 6, No. 3, pp. e17560.

Van der Perren, A., Toelen, J., Carlon, M., Van den Haute, C., Coun, F., Heeman, B., Reumers, V., Vandenberghe, L.H., Wilson, J.M., Debyser, Z. & Baekelandt, V. (2011) Efficient and stable transduction of dopaminergic neurons in rat substantia nigra by rAAV 2/1, 2/2, 2/5, 2/6.2, 2/7, 2/8 and 2/9. *Gene Ther.*, Vol. 18, No. 5, pp. 517-527.

Vande Velde, G., Rangarajan, J.R., Toelen, J., Dresselaers, T., Ibrahimi, A., Krylychkina, O., Vreys, R., Van der Linden, A., Maes, F., Debyser, Z., Himmelreich, U. & Baekelandt, V. (2011) Evaluation of the specificity and sensitivity of ferritin as an MRI reporter gene in the mouse brain using lentiviral and adeno-associated viral vectors. *Gene Ther.*, Vol. 18, No. 6, pp. 594-605.

Wong, L.F., Goodhead, L., Prat, C., Mitrophanous, K.A., Kingsman, S.M. & Mazarakis, N.D. (2006) Lentivirus-mediated gene transfer to the central nervous system: therapeutic and research applications. *Hum Gene Ther*, Vol.17, No. 1, pp.1-9.

Yi, Y., Noh, M.J., Lee, K.H. (2011) Current advances in retroviral gene therapy. *Curr Gene Ther.*, Vol. 11, No. 3, pp. 218-228.

Yizhar, O., Fenno, L.E., Davidson, T.J., Mogri, M. & Deisseroth, K. (2011) Optogenetics in neural systems. *Neuron*, Vol. 71, No. 1, pp. 9-34.

Zhang, F., Gradinaru, V., Adamantidis, A.R., Durand, R., Airan, R.D., de Lecea, L. & Deisseroth, K. (2010) Optogenetic interrogation of neural circuits: technology for probing mammalian brain structures. *Nat Protoc.*,Vol. 5, No. 3, pp.439-456.

6

Avian and Pandemic Influenza (API): A Prevention-Oriented Approach

Muhiuddin Haider and Jared Frank
School of Public Health, University of Maryland College Park, Maryland, USA

1. Introduction

The current public health approach to avian influenza focuses on control and management after an outbreak has already occurred. While control methods have numerous benefits and have played a large role in containing outbreaks, although not until some days, weeks, and/or months following the initial infection(s), this approach does have its limitations. Not only is a focus on control following an outbreak very costly, but during the time after initial infection and the subsequent spread, recent research has found issues with the control of diseased bird/poultry populations, the sanitary removal of the diseased carcasses of deceased birds/poultry, and the treatment of AI among infected humans. Through focus groups and observational studies, researchers have also found lapses in the training and outreach provided to those individuals most at risk of exposure and contamination by HPAI H5N1.

A prevention-oriented approach, one that utilizes multiple prevention-oriented measures, including the research of new vaccinations, the further development of health regulations and standards, a strong focus on the One Health Initiative, and improved training programs and awareness campaigns, should be utilized to address the issues associated with AI, especially bird-to-human transmission. This chapter aims to inform about the similarities and differences between control and prevention measures and the benefits of a prevention-oriented approach.

2. What is Avian Influenza?

Throughout the development of civilization, researchers have continuously discovered the means to prevent, treat, and/or control diseases that have plagued mankind. However, there are still many diseases, both those recently presented among human populations and some that have adversely affected health outcomes for years without effective treatment, that have yet to be successfully addressed and eradicated. AI, a relatively new emerging infectious disease of zoonotic origin, has caused upwards of tens of millions of deaths among domestic poultry and wild bird populations globally. Even more problematic is that numerous regions have seen scattered outbreaks, some with dozens of deaths recorded,

* The authors recognize the assistance of Zandra H. Andre

among human populations. Further, the disease continues to spread in spite of measures implemented by governments and various world health agencies to control such outbreaks.

AI, otherwise known as avian flu or bird flu, is an influenza type A virus. Although there are a variety of sub-types of influenza virus, due to the many combinations of the two components that comprise the virus, haemagglutinin (H) and neuraminidase (N), H5N1 is well known for its past and current virulence among both avian and human populations. As for the components of the virus, "haemagglutinin is a protein found on the surface of influenza viruses which is responsible for binding the virus to the cell that is being infected; neuraminidase is also found on the surface of influenza viruses" (FAO, 2011). In addition, when discussing the disease itself, there are two types of AI known as low pathogenic avian influenza (LPAI) and highly pathogenic avian influenza (HPAI). "While prevalent in many regions, LPAI poses very little danger to birds and almost zero threat to human populations" (Haider, Frank, & Noreen, 2010, p. 323). However, HPAI, the category in which H5N1 falls, has a much higher capacity for causing disease than LPAI, as shown by its virulence among birds and its impact on human health once transmitted.

2.1 History and transmission

Over the course of human experience with AI, as is the case with many other diseases, researchers have heavily documented all instances and factors surrounding outbreaks of the disease. Outbreaks among animals, transmission to humans, and negative effects on the environment have been recorded as to their duration, scale, and the total number of resulting cases and subsequent deaths among both birds and humans. In addition to these measureable aspects of outbreaks, researchers have also kept a comprehensive timeline of every case of animal and human infection as reported by countries across the world. From the initial isolation of the H5N1 virus in 1996 in the Guangdong Province of China to the recent outbreak of H5N1 in the Banteay Meanchey Province of Cambodia (September 12, 2011), the World Health Organization (WHO) has compiled a timeline, entitled *H5N1 avian influenza: Timeline of major events,* which consists of all the significant points in history to date regarding AI (WHO, 2011).

As mentioned, AI affects both domestic poultry and wild bird populations in many countries. Outbreaks, due to transmission between these various populations, can occur in the wild, at commercial farms and personal farms, and at live bird markets. Although everyone is at varying degrees of risk of contracting AI, there are particular actions that result in higher chances of infection. In fact, those who keep or sell poultry are at a high risk of bird-to-human transmission of AI. Specifically, "some of the factors that contribute to spread of HPAI from birds to humans include slaughtering poultry and preparing the meat in the home, direct contact with sick or infected birds, and the consumption of infected poultry" (Haider & Applebaum, 2011, p. 20). Often, many individuals lack education and training concerning the proper handling of poultry and the temperature at which meat should be cooked to prevent infection. As it relates to animal-to-human transmission, many are even unaware that they should minimize contact with the feathers, intestines, blood, saliva, and droppings of diseased or dead birds.

Unfortunately, not only are farmers and poultry workers contracting the disease through improper handling and cleaning of poultry, but, due to low hygiene conditions and close quarters, the disease continues to spread unchecked throughout domestic poultry and wild bird populations. If one domesticated bird has the strain, it can spread to other birds in the flock quickly, well before culling or other measures of containment can be effectively implemented. Fortunately, the virus has not yet mutated into a form that has high potential for human-to-human transmission.

2.2 Global impact of Avian Influenza

Along with the increasing numbers of deaths within domestic poultry and wild bird populations throughout numerous countries in the Middle East, Africa, and Asia, human deaths have also steadily continued to occur in spite of measures designed to control the spread of this disease. In fact, there are fifteen (15) countries that have reported human cases and/or deaths from HPAI H5N1. These countries are ordered from highest to lowest number of human cases/deaths through August 2, 2011 in Table 1:

Country of Impact	**Human Cases**	**Human Deaths**
Indonesia	178	146
Egypt	144	48
Vietnam	119	59
China	40	26
Thailand	25	17
Cambodia	16	14
Turkey	12	4
Azerbajin	8	5
Bangladesh	3	0
Iraq	3	2
Pakistan	3	1
Lao People's Democratic Republic	2	2
Djibouti	1	0
Myanmar	1	0
Nigeria	1	1

Data obtained from World Health Organization (WHO, 2011)

Table 1. Countries with HPAI H5N1 Human Cases & Deaths

As indicated in Table 1, AI has had a substantial impact on human populations in Southeast Asia (i.e. Indonesia and Vietnam) and an equally large effect in Northern Africa (i.e. Eygpt). Further, instances of the disease in such countries as Turkey, Iraq, Djibouti, and Nigeria indicate a potential trend in the spread of the disease further east through countries in both the Middle East and Africa. This potential trend may be further shown by the outbreaks of HPAI H5N1 among birds in countries including Israel, Japan, Republic of Korea, Mongolia, and the Palestinian Autonomous Territories [West Bank] (FAO, 2011). Should measures not be taken, as mentioned above, AI may continue to spread along this path through commonly utilized transportation routes for poultry, potentially threatening new, unsuspecting populations.

In their mid-year review of AI in 2011, the FAO found that within "countries in Asia and Africa—The People's Republic of China in East Asia, Vietnam in the Greater Mekong sub-region, Indonesia in Southeast Asia, Bangladesh and India in the Indo-Gangetic Plain, and Egypt in North Africa—H5N1 HPAI remains entrenched, and these countries are considered endemic for the disease" (FAO, 2011, p. ix). Three countries, Bangladesh in South Asia, Indonesia in South East Asia, and Egypt in the Middle East, have been increasingly representative of areas that have focused heavily on control methods.

In fact, as recently as June 22, 2011, The Ministry of Health of Egypt reported a human case of H5N1, which was confirmed by the "Egyptian sub-national laboratory for Influenza in Aswan and the Central Public Health Laboratories in Cairo, a National Influenza Centre of the WHO Global Influenza Surveillance Network" (WHO, 2011). In this instance, the individual developed symptoms, which were recognized, and they were placed in the care of a nearby hospital. While the individual received medical care, through treatment with oseltamivir, otherwise known as tamiflu, which slows the spread of the influenza virus between cells, there was a nine day delay between the development of symptoms and treatment, with death following a day after the first treatment. Unfortunately, it appears that in this case, as present in many communities that have little to no access to education or up-to-date medical facilities, that the symptoms were recognized much too late, the treatment was given long after symptoms were presented, and the exposure incident had not been fully discerned.

As another country which has been heavily impacted by HPAI H5N1, Indonesia currently has the most recorded human cases and deaths from the disease and thereby serves as a very important area in terms of controlling and/or preventing disease. Unfortunately, problems such as "the complex and weakly regulated structure of the poultry sector has hampered the control and prevention of avian influenza in Indonesia" (FAO, 2011, p. 66). After years of trying to control outbreaks on a case-by-case basis, the FAO along with the Ministry of Agriculture of Indonesia and private sector market-traders have begun to team up to implement prevention-oriented measures. A specific example includes initiating "a cleaning and disinfection program for poultry transport vehicles at collector yards" (FAO, 2011). Instead of having to cull large numbers of poultry to control the spread of the disease and subsequently losing money, preventive measures are beginning to be recognized as both necessary and economical. In case of an outbreak, the FAO Emergency Centre for Transboundary Animal Diseases' Avian Influenza Programme has provided funding to help the government of Indonesia "to implement a host of avian influenza prevention, surveillance, response, control, research and communication activities nationwide" (FAO, 2010).

Bangladesh is also seen as a hotspot for AI. To date, there has been much done by the international community in the way of supporting the government of Bangladesh to control outbreaks of HPAI H5N1.

In order to further emphasize the global impact of HPAI H5N1, Figure 1 below gives a global geographic representation of the spread of AI over the past eight years:

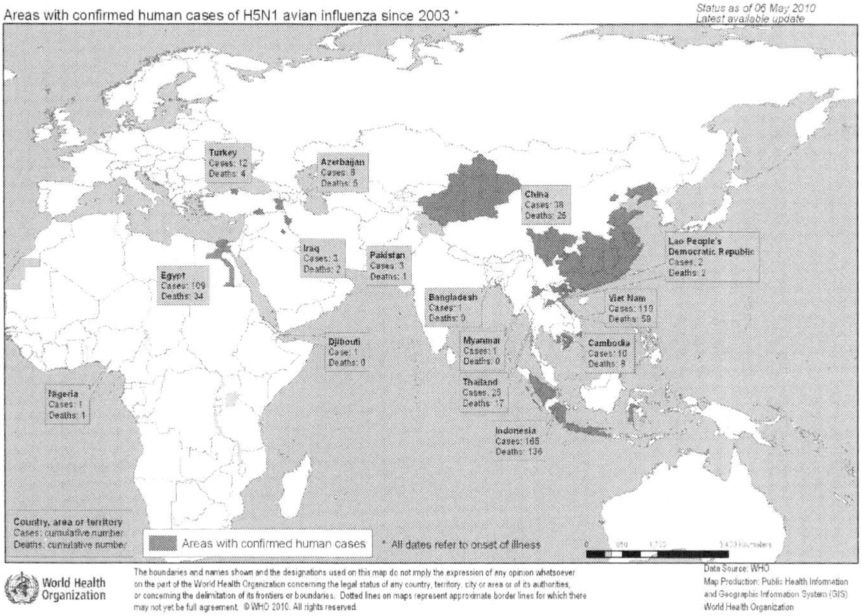

As seen in Figure 1, the areas colored in red on the map are those areas with confirmed human cases of H5N1 since 2003, as of the date May 6, 2011 (WHO, 2011). The number of deaths attributed to the disease among humans within each country at this particular point in time has also been provided.

Fig. 1. Areas with confirmed human cases of H5N1 avian influenza since 2003*

While the tables and figures above have examined the distribution of cases and deaths by country, Figure 2 gives a total of the same information for the years 2003 through the present:

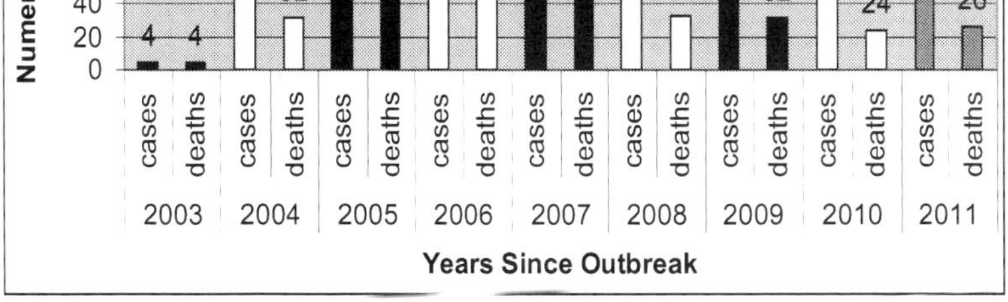

Fig. 2. HPAI H5N1 Cases & Deaths

Based on Figure 2, following a rapid increase in the number of human cases and deaths over the first few years of AI's reemergence, the numbers seem to have since declined. However, regardless of any decline in HPAI H5N1 human cases and deaths, the fact remains that, as reported by the WHO as of August 2, 2011, there have been approximately 566 human cases of AI globally, with 332 total deaths (WHO, 2011).

2.3 Current status & future risk

Presently, due to widespread outbreaks of AI among poultry, especially within numerous developing countries, there has been a large focus on the utilization of control methods, such as through the culling and/or quarantine of diseased birds. Unfortunately, as will be further examined in this chapter, methods of control are often expensive, countries lack up-to-date training and facilities, and resources are often unavailable as needed to control outbreaks. Instead, the authors believe a prevention-oriented approach can reduce the impact of AI going forward. In fact, through the utilization of mass education, development and implementation of regulations provision of vaccines, and a collaborative approach taken through the One Health Initiative, we believe prevention, as compared to control, has a major role to play in combating this disease. However, we are not so quick as to dismiss the benefit of control methods. As such, current control methods will be examined, followed by a comparison of control and preventive methods, and then a thorough review of effective prevention methods.

3. Control of Avian Influenza

In an effort to deal with outbreaks of AI, health organizations, governments, and local and state agencies within numerous countries have been working to develop and implement control measures to prevent further loss due to the disease. Currently, there is no treatment for AI among humans beyond treating the symptoms, such as a cough, fever, or trouble breathing, associated with the disease. Unfortunately, animals that become infected with the disease are not so treatable. In fact, "control measures such as rapid culling, extensive quarantines, and sanitary measures have been taken" (Haider et al., 2010, p. 324). Additional methods for control include the rapid removal of the fecal matter, blood, and other bodily fluids from diseased/infected birds so as to prevent runoff. These measures, however, are not considered truly effective as the quarantined birds often die, the culling must encompass some not yet diseased birds because they may carry the virus and infect others, and any sanitary measures taken may not encompass all areas where the outbreak occurred. However, the FAO has reported that "good husbandry, proper nutrition, and broad spectrum antibiotics may reduce losses from secondary infections" (FAO, 2011).

Recently, the FAO has recognized that implementation of various control measures "based around early detection and stamping out, appears to have reduced the number of cases but the virus has not been eliminated" (FAO, 2011, p. 51). Concerning potential issues with control, the FAO has recognized various constraints for disease control and responses (FAO, 2011, p. 55):

- The requirement of a government order before culling delays early response

- The quality of post-outbreak disposal, cleaning and disinfection is substandard
- The location of commercial farms in densely populated areas makes movement control of people and animals difficult
- Borders are long and porous
- The source of infection is unknown
- Tracing infection forward or backward is difficult

On the other side of the argument, however, it is noted that these control measures have resulted in a decrease in the time required for detection, laboratory testing, and culling (i.e. stamping out) after infection. Regardless, the focus on AI has seen a shift in methods towards a preventive approach. Currently, regulations are being developed, such as those originating from the Animal Slaughter and Meat Quality Control Act of 2010, and education/training is being provided to help ensure faster diagnosis of the disease so control measures can be implemented.

As of 2007, the Third Global UNSIC-World Bank (pg. 67) Progress Report found overall improvement in the ability of numerous countries to respond to HPAI outbreaks by way of new surveillance techniques, updated laboratory systems, and an increased capacity of health systems, though varying by region, to handle potential and real threats to human health. However, this same report found multiple issues such as an "insufficient joint working of animal and human health surveillance and response networks" and lack of significant translation into behavior change among individuals and communities of the "increasing awareness of the threat posed by HPAI H5N1" (UNSIC & World Bank, 2008, p. 67). With proven measures of control already in place, future efforts must now be placed on the development, build up, and implementation of prevention-oriented measures.

3.1 Costs of control

Since the outbreak of HPAI H5N1 in 1996 and the subsequent outbreaks across the world, billions of dollars have been pledged and globally distributed by numerous countries in an effort to control and prevent the rapid spread of this disease. As noted by the United Nations (UN) System Influenza Coordinator & World Bank, in their document, *Responses to Avian Influenza and State of Pandemic Readiness,* the rapid spread of HPAI has resulted in "significant socioeconomic losses, numerous human deaths, and the potential threat of a human pandemic influenza" (UNSIC & World Bank, 2008, p. 8).

There have been many advances in control measures such as the development of surveillance techniques, updating health care facilities, such as hospitals and laboratories, and improving upon health systems and outbreak response networks. In fact, the ability of a country to carry out effective disease management in cases of an outbreak is highly dependent upon such constructs as capacity building, the training of workers and volunteers and provision of commodity supplies, and infrastructure development. Further, logistics management focuses on "issues pertinent to implementing disease management strategies such as space and equipment availability, staffing and human resource skills, supplies of relevant commodities, recordkeeping and reporting, and transportation" (Haider & Applebaum, 2011, p. 9). Additionally, there is an ongoing need for the provision of clean water and cleaning supplies, such as sprayers, detergents, masks, and gloves, especially in live bird markets, and ongoing surveillance/monitoring. As expected, however, there are high costs associated with infrastructure development, especially in (low income) developing countries, and training of staff and volunteers to successfully implement biosecurity measures and carry out an effective outbreak response.

Though government, and international, assistance is often provided in cases of an outbreak, it is incumbent upon each country to develop, fund, and implement their own disease management programs. As an example, the United States Agency for International Development's (USAID) Stamping Out Pandemic and Avian Influenza (STOP AI) provides assistance, such as through training and public-private partnership development, in order to better prepare countries for responding to and recovering from HPAI H5N1 outbreaks. Along with providing technical assistance, STOP AI "aims to mobilize public and private sector partners as well as NGOs to implement systematic and sustained behavioral changes that will result in measureable improvements in biosecurity" (Haider & Applebaum, 2011, p. 6). However, such collaborative efforts are often limited by issues including opposing political views and the time involved concerning the political process, private sector motivation (i.e. direct benefit), public sector oversight and regulation, and cost-sharing. Going forward, it is important to note that efforts involving capacity building, logistics management, and communication for behavior change are not only measures used to manage (control) and/or eradicate AI in live bird populations, but can play a significant part in outbreak, and subsequently disease, prevention (Haider & Applebaum, 2011, p 8).

As noted above, while there are numerous methods for controlling the spread of influenza, the costs associated with control of an outbreak, such as rapid mobilization of a global health response, mass production of treatments and other health services, and related efforts, are very high. However, a focus on prevention has shown a decrease in the overall cost associated with disease. Concerning past experience with multiple outbreaks of influenza, though not on the scale of a potential outbreak of AI at the level of human-to-human transmission, the "WHO recommends annual immunization of at-risk persons as the best and most cost-effective strategy for reducing influenza-related morbidity and mortality" (WHO, 2011). Similarly, on a more basic level, the benefits received and costs avoided as a result of strong hygiene practices cannot be overlooked. "Of these practices, hand hygiene and surface cleaning are among the simplest and most cost-effective ways to prevent transmission of the highly pathogenic avian influenza virus" (WHO, 2011).

4. Control & prevention

Although control and prevention appear to be two different methods for addressing the same problem, it should be noted that researchers, agencies, and governments can approach AI successfully utilizing measures from both areas simultaneously. In fact, the FAO Emergency Centre for Transboundary Animal Diseases (ECTAD) produced a document in January 2007 called *Protect Poultry – Protect People: Basic advice for stopping the spread of avian influenza*. This document provides a quick, but thorough, review of current prevention and control measures when dealing with potential and real cases of AI among wild birds, domestic poultry, and humans, and how transmission may be prevented between animal-to-animal and animal-to-human (ECTAD, 2007). Further, Figure 3 below examines measures of control and prevention and the overlap between both approaches.

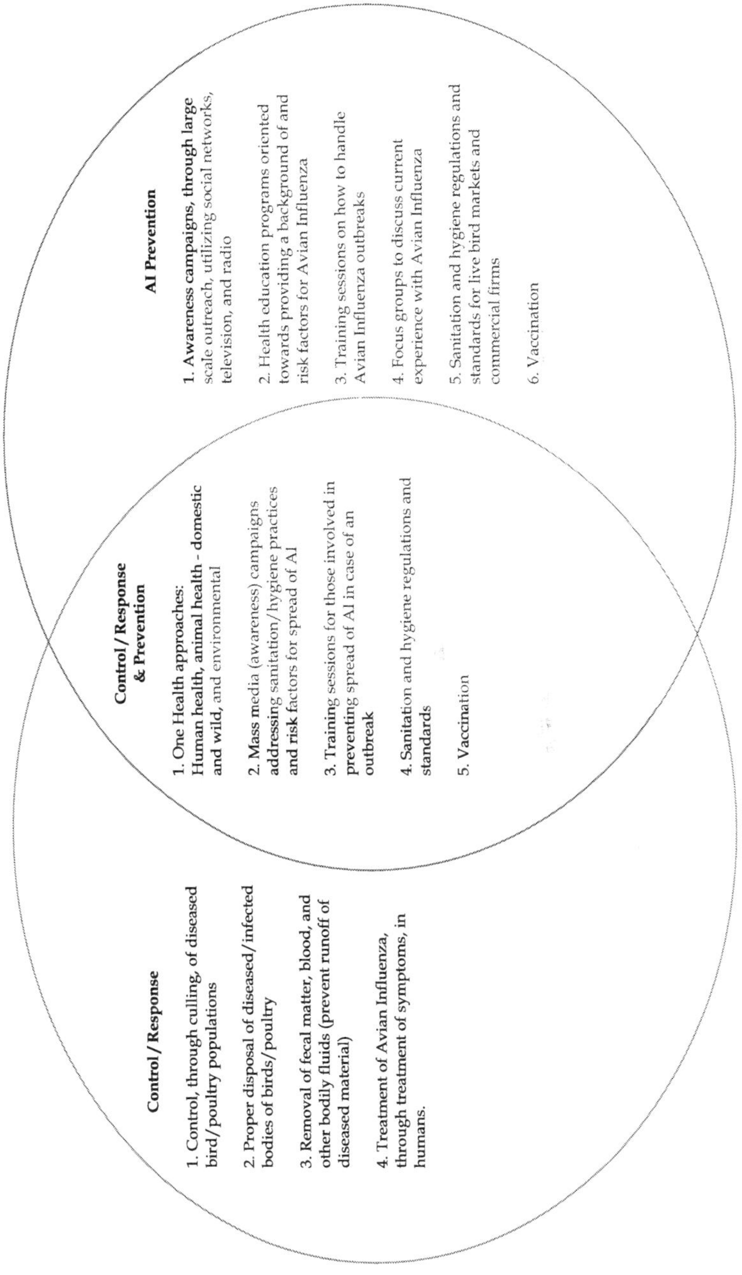

Fig. 3. Control vs. Prevention: Compare & Contrast

5. Prevention of Avian Influenza

Although control and prevention both have a role in dealing with AI, as shown in Figure 3, the majority of the focus should be placed on prevention. While control aims to prevent further spread of a disease in a region in which an outbreak has already occurred, prevention seeks to address potential vulnerabilities ahead of time in order to prevent an outbreak from occurring in the first place.

As seen by the examination of past and present control measures utilized for handling AI outbreaks, it is necessary to implement new strategies to address the issues faced as a result of this disease. Recent research has shown that it is possible to utilize resources more strategically by adopting a prevention-centered approach, especially in those areas currently at a high risk of an outbreak of AI. In fact, to date there have been lapses in training and outreach, inadequate education, and poor dissemination of health information. Researchers have also noted that "overall, knowledge is necessary, but NOT sufficient to produce behavior change. Perceptions, motivation, skills, and factors in the social environment play important roles" (Glanz & Rimer, 1995). Therefore, education programs and interventions going forward should incorporate comprehensive health education with related behavior change methods so that individuals are not only knowledgeable about what needs to be changed, but they can also have the self efficacy to carry out the healthy behaviors.

As it pertains to defining a country's risk status, all countries are at varying levels of risk of AI due to reasons including improper procedures and lack of food and agriculture standards. High risk areas may be defined as those that are currently experiencing or have already experienced an outbreak. Similar to the manner in which control measures are currently being implemented to combat AI, potential preventive measures can also be implemented and then analyzed as to their effectiveness. If successful interventions can be found to produce significant results, the intent would then be to widely disseminate these measures in other high risk areas, areas considered to be at low risk (those which have had limited outbreaks of the disease), and even in those regions that have had no previous experience with AI, so as to prevent the possibility of a future outbreak.

The prevention of AI can be defined as hindering both the outbreak and spread of the disease. As seen by the various preventive measures to be examined, there will be a focus placed on the Social Ecological Model of Prevention, developed by the Centers for Disease Control and Prevention (CDC), which aims for better health outcomes through a tiered approach, at the individual, community, and societal levels, for identifying areas for prevention activities. This model can be best characterized as an 'upstream approach' where, as one moves upstream, one makes their way to larger streams, which then lead to rivers, and ultimately, to the source of the problem.

This is a proactive, systemic approach, though fairly complex, which better enables researchers to get to the core of the problem. The overall aim is to reduce morbidity and mortality rates. By preventing an outbreak before it occurs, resources can be directed toward broader improvements in sanitation and hygiene practices that will positively affect not only AI control efforts, but those of many other communicable diseases.

5.1 Awareness campaigns & training sessions

One of the main prevention oriented efforts currently in place in many countries and utilized by world organizations is the mass utilization of education programs through awareness campaigns and targeted training sessions. The idea behind the use of education programs, whether at the individual, group, or community level, is to provide information on handling the spread of AI, especially concerning how the outbreak of the disease can be prevented in the first place. Research indicates that the aforementioned awareness campaigns, including targeted training sessions and health information dissemination, should aim to prevent AI by focusing on a variety of aspects including how to:

- Maintain a sanitary environment
- Maintain the overall health of poultry/livestock on the farm
- Properly handle diseased birds
- Properly dispose of the diseased bodies of dead birds
- Properly cleanse birds/poultry and cook the meat appropriately before consumption
- Transport poultry/livestock under conditions which prevent the spread of disease in case of an outbreak

While these methods are very helpful, there are additional measures that may be taken by farmers and commercial businesses to prevent virus transmission, and resulting issues, among their poultry and other at risk populations. These additional methods include meeting the regulations and standards set forth by local, state, and federal governments and the provision of vaccinations.

The manner in which education is provided is crucial and depends on the audience of interest. The primary way in which information should be disseminated, which could reach a wider audience and in a more cost effective manner, is through awareness campaigns (media) by way of television, radio, and social networks, where available. Through these methods, information can be provided on sanitation and outbreak control measures, new policies and regulations, and vaccine dissemination. For a more direct approach, community oriented training sessions would educate the population at a more personal level and potentially result in greater behavior change as the community works together to ensure prevention. Key limitations of these methods include availability, especially in the developing countries, and cost.

When considering these widespread education programs, the focus should primarily be on training farmers since flu transmission occurs primarily during bird/poultry handling and cleansing process. If the lay population comes into contact with avian flu, the interventions then become control-oriented. However, in addition to farmers, training sessions and education-oriented campaigns should target both men and women, especially the wives of the farmers. Although the most likely recipients of this information would be [male] farmers, recent focus groups among women in Pakistan find that, "to ensure a faster spread of awareness of Avian Influenza throughout the various regions, training sessions and related health education programs should focus on incorporating both genders" (Haider, Frank, & Noreen, 2010, p. 329). As Haider et al. discovered, after conducting focus groups with two groups of women in Pakistan, many of whom had husbands whose occupation was farming and/or had knowledge about AI through their husbands going regularly to live bird markets, it was apparent that there was a lack of education within the various communities in the region of study. As a result of this study, "more training sessions should be scheduled, and the information covered should be more extensive, especially regarding the transmission of the virus and the most recent preventative measures" (Haider et al., 2010, p. 332).

5.2 Development & provision of vaccinations

The utilization of educational outreach and training sessions covers one aspect of the prevention-centered approach. However, an integral part of these awareness and information driven campaigns is the focus on spreading awareness of current AI vaccination measures. Specifically, the method of vaccination has been proven to be an effective tool for preventing disease by enabling resistance, thereby reducing the chance of becoming infected during an outbreak. Similarly, vaccination of poultry and humans helps to control disease in both populations by stalling or stopping the spread of the disease in question. In fact, "in developing countries, vaccination programmes in avian species have been recommended recently, however it will require concurrent management of local husbandry practices and industry compliance to eradicate the disease rather than the establishment of an endemic situation" (Capua, 2007, p. 5645). While developed countries have had lower levels of risk of an outbreak of AI, as indicated by both the lack of animal and human cases and deaths from the disease and the architecture currently in place, including regulations and standards, health education programs, and health services, many of these countries are not taking any chances. "In developed countries vaccination is being used as a means of increasing resistance of susceptible animals to reduce the risk of introduction from the reservoir host or to reduce secondary spread in densely populated poultry areas" (Capua, 2007, p. 5645).

Officials, researchers, and businessmen/farmers, are aware that control measures such as depopulation through massive culling is not feasible when both food supplies and economic stability are at risk of upset due to the spread of disease. By focusing on vaccinations, the health and safety of domestic poultry can be maintained as long as the efficacy of the vaccine is constantly monitored, infected birds are managed appropriately to prevent infection of healthy populations, and most importantly, regulations and standards are met and exceeded such that HPAI H5N1 is prevented from gaining a foothold in any region. Of course, some major questions still need to be answered concerning the utilization of vaccines. As mentioned above, the current and future effectiveness of vaccine use is reliant on monitoring and continuous study. Another potential problem is the question of whether the vaccine will be developed, produced, and distributed by a government organization or private entities, specifically who will be responsible for the provision of the vaccine. Finally, an issue that still needs to be addressed is that regardless of the research showing prevention to be far less expensive than control, the mass production and distribution of vaccines is very costly for both the developer/manufacturer and the individual businesses and farmers aiming to prevent any outbreaks of the disease by vaccinating themselves and their populations of livestock. A more in-depth analysis of cost control measures and the additional resources necessary for successful implementation of vaccine campaigns is especially warranted.

5.3 Policy & regulations

Another important aspect of prevention is the development of regulations. These regulations should foremost address sanitation and hygiene among live bird markets and commercial businesses in an effort to prevent and control future outbreaks. By setting forth responsibilities for the applicable parties to follow, governments aim to maintain the safety, well-being, and overall health of a population. As an example, for areas such as the United States which have had no first-hand experience with AI, the United States Department of Agriculture (USDA), in an interim rule, prohibited the import of wild birds/poultry and poultry products (i.e. meat) from any and all regions where a subtype of HPAI, such as H5N1, exists.

Unlike in developed countries, which have a modicum of control over outbreaks of AI and related diseases due to health regulations and significant oversight of the food production industry, developing countries have a different set of concerns. Specifically, regulations that should receive focus in these countries include those that better protect the food protection process such as processes for handling and transporting poultry products and registering the farms responsible for both of these tasks. Currently, the FAO, as part of their Animal Production and Health Paper, *Approaches to controlling, preventing, and eliminating H5N1 Highly Pathogenic Avian Influenza in endemic countries*, notes key areas, specifically biosecurity measures, which need to be addressed through regulations for the small commercial businesses and personal farms that are found throughout those countries affected by AI. The regulations should address the fact that (FAO, 2011):

- Live bird markets need to be improved
- Poultry slaughter houses need to be modernized
- Decontamination procedures are not sufficiently monitored
- Infrastructure in poultry production needs to be upgraded including building poultry processing plants
- Limited integration of small-scale farming into larger corporate farming enterprises
- Animal Slaughter and Quality Control Act will be revised but uncertainty around its implementation
- Existing regulations are not well executed or enforced

Sanitation is one of the most important aspects of maintaining the health of a large animal population in close quarters, which in this case refers to domestic poultry. Infrastructure, specifically poultry slaughter houses and live bird markets, needs to be updated in terms of procedures, tools, and oversight. Furthermore, in case of an outbreak, "there should also be a standardisation of reporting procedures for outbreaks of disease to internationally accepted standards" (The World Bank, 2006). The World Bank notes that the development of an operational manual would not be unwarranted, one that addresses 1) how to react in the event an outbreak occurs, 2) which individuals, organizations, and/or entities are responsible for certain tasks to control further outbreak or response to human infections, and 3) how will coordination be handled. By ensuring the health of the environment and sanitary living conditions for the poultry under care, especially through well thought out, approved regulations, there is a better chance that the H5N1 virus may be resisted and/or the spread and impact of the virus is decreased.

5.4 The One Health Initiative

The above mentioned prevention strategies are key components of preventing the outbreak and spread of AI. However, these efforts need to be utilized together in order to increase their effectiveness. Specifically, these prevention strategies should be incorporated into the One Health Initiative. This initiative, which is carried out in collaboration by many agencies including the American Medical Association (AMA), the CDC, and the USDA, is "dedicated to improving the lives of all species—human and animal—through the integration of human medicine, veterinary medicine and environmental science" (One Health Initiative, 2011). The One Health Initiative, which is a concept aimed at increasing interdisciplinary collaboration "in all aspects of health care for humans, animals and the environment" is based on the premise that the health of these three areas as a whole are linked, whereby a detrimental effect on the health outcomes of one area could have negative outcomes for that of another.

In order to truly push AI prevention measures into the realm of One Health, some components need to be addressed. A major component to be expanded upon, as mentioned above, is comprised of a dual focus on both human and animal health research, such as through the development of new vaccinations and antiviral drugs, including the promotion of alternative administration measures to make widespread vaccination feasible within each population. Another component of interest is the improvement upon these methods and the addition of new human health behavior interventions focused at the individual, community, and population levels. Furthermore, enhanced wildlife surveillance techniques along with additional funding for such endeavors would serve to prevent the spread of infection along migratory paths. As noted, the migration of wild birds can quickly and efficiently spread disease from one region to the next if the proper steps are not taken immediately, assuming an availability of the appropriate resources. Another key issue addressed by the One Health Initiative is the increased involvement of animal health professionals, including veterinarians, alongside those physicians, scientists, etc. focused on human health. Finally, as discussed under the methods for prevention and control of the disease, there should be more of an emphasis placed on ecosystem health. This includes such aspects as proper disposal of carcasses to maintain clean water supplies, sanitary living conditions for humans and animals, and sustainable production of poultry.

Efforts to support a prevention-centered approach to pandemic AI, particularly HPAI H5N1, can be used to strengthen a nation's public health infrastructure, which will ultimately result in greater public health and security gains than any reactionary response could possibly hope to address. While the necessity has previously been to control and mitigate outbreaks once they have occurred, we are in a strategic position to move towards a more sustainable focus on prevention. It is time to seize the opportunity to get ahead of the threat and to focus resources on stopping outbreaks before they occur. Success can no longer be measured strictly by the control of outbreaks and minimization of the associated human cases and deaths, but by the ability of each and every country to save the lives of animals and humans alike through an effective, efficient, prevention-centered approach.

6. The innovation of a prevention-oriented approach

Instead of predominantly focusing on containment, through quarantine and culling, or related control measures, countries should implement prevention-oriented methods in regards to HPAI H5N1. By preventing future outbreaks, not only will we see a reduction in the number of deaths of both humans and animals, but the economic burden of many countries and health organizations can be decreased.

The successful implementation of a prevention-oriented approach, one that has a comprehensive, and targeted, focus on prevention, would have the potential for measureable gains in the struggle against AI. As such, an approach which addresses the aforementioned major areas of prevention, specifically education and mass awareness, widespread vaccination campaigns, implementation of regulations on private sector sanitation and hygiene, and the One Health Initiative, has not yet been seen. Further, once proven control measures, such as targeted education and training sessions and the provision of antibiotics and related commodities in response to outbreaks, have been incorporated into this approach, the potential exists for reducing the impact of HPAI H5N1 worldwide. However, to date, there has not been the availability of, or funding for, such an all encompassing approach. In regards to many developing countries, and even some developed countries, another question concerns the diffusion of the numerous aspects of this approach, assuming funding, approval, and subsequent implementation.

As it relates, FOMENT, a communications tool developed by Dr. Muhiuddin Haider to compliment the well known and often utilized Diffusion of Innovations Theory (DOI), should be employed in this case. In order for this prevention-oriented approach to be effective, all aspects must be adopted and successfully implemented from the government level down to the individual level. FOMENT includes six components, which complement those of DOI, that could be effectively utilized to encourage the adoption of such an approach. Specifically, as noted by Dr. Haider, FOMENT consists of the *Focus* on a specific behavior change, the *Organization* of the behavior change program, *Management* which supports and approves of the behavior change plan, an *Environment* which is conducive to behavior change, a *Network* in which to diffuse innovations at both individual and organizational levels, and the *Technology* available to diffuse innovations (Haider, Pal, & Al-Shoura, 2005). FOMENT offers a broad view of the various change agents, the advocate for the innovation and the potential adopters, the relationships between and environment for all interactions, and especially the technology that may facilitate or hinder successful implementation (i.e. diffusion).

7. Determination / Conclusion

Without sufficient education, awareness campaigns, and targeted training sessions, including the promotion of behavior change, many individuals will remain unaware of sanitary ways to handle and transport poultry and how to properly cook meat. Similarly, vaccination campaigns for both animals and humans can prevent at-risk populations from contracting the disease through animal-to-animal or animal-to-human transmission. The development of regulations will also serve to set standards that must be met concerning quality of health and the provision of a sanitary environment. By acting upon the aforementioned methods to prevent disease, especially as part of the One Health Initiative, by focusing on animal, human, and environmental health in all aspects of addressing AI, the world as a whole may indeed see the end of AI.

Following the successful implementation of a prevention-centered approach, a major goal is making prevention 'sustainable'. Key goals following the utilization of preventive measures include the retention of lessons learned from the training provided, the continuation of processes in place, and the adherence to rules and regulations. Specifically, training programs should be maintained going forward, especially in cases where H5N1 appears to have been nearly eradicated. Health education programs should be continued and expanded upon regarding AI response and prevention information, and the education given should be as up-to-date as possible. Concerning regulations, butchers and farmers who prepare their own meat receive education and farms and slaughter houses should be inspected for sanitation and proper disposal. Hygiene regulations should be increased and there should be properly enforced penalties for those who violate proper hygiene in regards to handling infected poultry along with incentives for those who maintain/exceed standards. Furthermore, assistance should be provided to the governments to potentially help those who rely on poultry as a source of income so that they do not need to sell infected meat in desperation and will dispose of it properly without worrying about losing a significant amount of money, thus ensuring adherence to the regulations set forth. In the end, the need for clinical management of H5N1 may be decreased and potentially eliminated by managing H5N1 ahead of an outbreak, especially through a focus on live birds. Each country should ensure they are working on increasing awareness and informing the population of the significance of the disease, especially through working with the government to implement more programs, providing public service announcements, and particularly by education villages/communities.

As we have seen, when outbreaks occur, countries want to control and respond on an emergency basis, thus using more resources than would have been necessary had the outbreak been prevented. By taking a proactive, instead of a reactive, approach towards AI, countries will see less utilization of essential resources, create more awareness, which leads to a better educated populace, and ultimately better control and prevent the spread of AI.

8. References

Capua, I. & Marangon, S. (2007). Control and prevention of avian influenza in an evolving scenario. *Vaccine*. *25*(30): 5645-5652.

FAO. 2011. *Approaches to controlling, preventing and eliminating H5N1 Highly Pathogenic Avian Influenza in endemic countries.* Animal Production and Health Paper. No. 171. Rome.

FAO Emergency Centre for Transboundary Animal Diseases (ECTAD). (2007, January). Protect people - protect poultry: Basic advice for stopping the spread of avian influenza. Retrieved from http://www.fao.org/docs/eims/upload//207623/FAO_HPAI_messages.pdf

Food and Agriculture Organization of the United Nations (FAO). (2010). FAO at work 2009-2010: Growing food for nine billion. Retrieved from http://www.fao.org/docrep/013/am023e/am023e00.pdf

Food and Agriculture Organization of the United Nations (FAO). (2011). *Avian influenza: Background.* Retrieved from http://www.fao.org/avianflu/en/background.html

Food and Agriculture Organization of the United Nations (FAO). (2011). *Avian influenza in 2011: Mid-year review.* Retrieved from http://www.fao.org/avianflu/en/news/ai_review.html

Glanz, K. & Rimer, B. K. (1995). *Theory at a Glance: a Guide for Health Promotion Practice.* NIH Pub # 97-3896, Bethesda, MD: National Cancer Institute.

Haider, M. & Applebaum, B. (2011). Disease Management of Avian Influenza H5N1 in Bangladesh: A Focus on Maintaining Healthy Live Birds. In *Health Management.* Croatia: InTech Open Access Publisher.

Haider, M., Frank, J., & Noreen, S. (2010). Analysis of avian influenza with special focus on Pakistan. *Journal of Health Communication. 15*(3): 322-333.

Haider, M., Pal, R., & Al-Shoura, S. (2005). Diffusion of innovations and FOMENT: A synergistic theoretical framework in health communication. In. M. Haider (Ed.), Global public health communication: Challenges, perspectives, and strategies (pp. 1-24). Sudbury, MA: Jones and Bartlett Publishers.

Noreen, S. (2008). Report on level of knowledge of rural women about Avian Influenza through focus group discussion (pp. 1–18). Peshawar, Pakistan: Ministry of Agriculture, Livestock, and Cooperative Department, Government of NWFP.

One Health Initiative. (2011). *One health initiative: Mission statement.* Retrieved from http://www.onehealthinitiative.com/mission.php

The World Bank. (2006, May). *World Bank will support Bangladesh's avian flu program.* Retrieved from http://www.worldbank.org.bd/WBSITE/EXTERNAL/COUNTRIES/SOUTHASIAEXT/BANGLADESHEXTN/0,,contentMDK:20936997~pagePK:141137~piPK:141127~theSitePK:295760,00.html

UN System Influenza Coordinator & World Bank. 2008. *Responses to avian influenza and state of pandemic readiness.* Fourth Global Progress Report. New York.

World Health Organization (WHO). (2010, May 6). *Areas with confirmed human cases of H5N1 avian influenza since 2003*.* Retrieved from http://reliefweb.int/sites/reliefweb.int/files/resources/A724B3AD85DDDF2C852577230060879A-map.pdf

World Health Organization (WHO). (2011). *Avian influenza: food safety issues.* Retrieved from http://www.who.int/foodsafety/micro/avian/en/index1.html

World Health Organization (WHO). (2011, September 12). *H5N1 avian influenza: Timeline of major events.* Retrieved from http://www.who.int/influenza/human_animal_interface/avian_influenza/H5N1_avian_influenza_update.pdf

World Health Organization (WHO). (2011, August 2). *Global alert and response (GAR): Cumulative number of confirmed human cases of avian influenza A (H5N1) reported to WHO*. Retrieved from http://www.who.int/influenza/human_animal_interface/EN_GIP_LatestCumulativeNumberH5N1cases.pdf

7

The Role of 'Belladonna 200' in the Prevention of Japanese Encephalitis (JE) Virus Infection

Bhaswati Bandyopadhyay et al.*
School of Tropical Medicine, Kolkata
India

1. Introduction

With the advent of modified improved JE vaccines, JE preventive program is now going on in full swing throughout the globe, particularly in South-East Asian countries where the disease is highly prevalent. However, recent unswerving studies indicated that we should not satisfy ourselves with the belief that we could be able to stop the spread of this dreaded disease in near future. This is because as much as forty per cent infected children was found vaccinated questioning the efficacy of the vaccines itself, which are currently in use and frequent failures to access the deep rural areas, and in the vast forest areas by the dedicated trained vaccination team due to uninterrupted seasonal and other natural catastrophes which becomes a major setback in the program. Thus although almost all developing countries in the endemic zones have accepted the control program by vaccination, but the outcome of such an elaborate and difficult protocol appears gloomy. It is true that a speculative encouraging approach may not be practically feasible in a wide scale vaccination program considering the unreachable vast tropical rural areas and taming a gigantic inadequately educated population from aversion in participation in such a vaccination program.

There are also failures of vaccination of the JE reservoirs too. Thus it becomes a worthless approach when we try to vaccinate a pig population due to their extremely rapid growing population in a herd. Controlling the vectors is also an uphill task which appears totally frustrating due to difficult natural scenario of the vector breeding areas in the endemic zones.

All these aspects straightforwardly indicate that we should search all other possibilities in controlling this problematic disease taking a toll of young lives and creating a sizable disabled population in the society. In such an attempt we find some evidences that a homeopathic medicine "*Belladonna*" is used by some homeopathic practitioners in individuals to protect them from JE. Thus this paper contains details of our study to find out if there is any real protective role of this medicine in JE. This experiment was done in two different models - JE pock formation on chick chorioallantoic membrane (CAM) and JE infection of suckling mice by intracerebral route where mother of the litter was treated with

---

* Satadal Das[2], Milan Sengupta[1], Chandan Saha[1], Chinta Raveendar[2], Rathin Chakravarty[2], Chaturbhuja Nayak[2] Anil Khurana[2] and Krishnangshu Ray[1]
[1]*School of Tropical Medicine, Kolkata, India*
[2]*CCRH, Department of AYUSH, Government of India, New Delhi, India*

the medicine. This chapter contains details of these studies where we could find a bright outlook of such a preventive measure. This chapter also contains details of the procedures regarding studies of JE virus on CAM which is not available in details elsewhere in any published text books or journals, as this was standardized by us, and similarly although JE virus infection in suckling mice is well known but formulating a protocol of the study on prevention by treating mothers is a new approach. Thus these methodology details will be helpful for those who are working in this field or are going to learn more regarding these aspects in future.

2. Materials and methods

2.1 Place of study

This chapter is based on the study conducted at the Virology unit, Department of Microbiology, School of Tropical Medicine, Kolkata.

2.2 Mice animal model

The most common choice of host for isolating arboviruses is still the suckling mice. In this experiment, Swiss albino mice, Webster strain were used after obtaining permission from the Ethical Committee of the Institute. The animals were maintained in the mice colony of School of Tropical Medicine, Kolkata.

2.3 The JE virus

In this experiment the virulent Nakayama strain (Source human, year 1935, location Japan, GenBank accession no. EF571853, Genotype III) was used.

2.4 The medicine

Belladonna is a well known medicine used by homeopathic practitioners in symptoms of encephalitis. Thus Raue described as early as 1885 that Belladonna is indicated in diseases of nerve and brain characterized by the violent headache, drowsiness and delirium; dilated pupils; double sight (Raue 1885, reprinted 1975). There are different Belladonna preparations available in the market but in this study we used aqueous preparation of Belladonna 200 which was procured commercially. The medicine was prepared according to standard procedures advocated by Homeopathic pharmacopoeia of India (Ministry of Health, Government of India, 1971,1:1,7-16,72). Initially we started experiment with Belladonna 6 and although average survival time of the suckling mice of treated mothers after inoculation of the virus was increased from 36 h in controls to 50 h, but all the mice died (unpublished data). Later we found that C. V. Boenninghausen, a veterinary doctor as well as a renowned homeopathic practitioner described in 1843 that Belladonna 200 is the ideal medicine in experiments with mice. Thus we used Belladonna 200 in all successive lots in this experiment.

2.5 Egg inoculations (Bandyopadhyay et al, 2010)

For preventive studies one dose (50µl) of the aqueous dilution of the selected medicine was inoculated in the chorioallantoic membrane followed by the administration of 50 µL of the

JE viral suspension, 5-10 minutes later. The chorioallantoic membrane consists of an outer layer of stratified epithelium which constitutes the respiratory surface of the egg and an inner layer of endoderm (the lining of the allantoic cavity).

Dermotropic viruses (poxviruses and some herpes viruses) and JE viruses grow on this membrane, and at low concentrations, they produce discrete foci of cell proliferation and necrosis (pocks). The membrane was therefore used to assay JE viruses in this study. Different viruses cause pocks of different color and morphology, and this is also of diagnostic value for distinguishing between different viruses. One day old fertile hen's (White leghorn) eggs were obtained from State Poultry Farm of Govt. of West Bengal, Tollygunge, Kolkata. They were collected from healthy flocks which were maintained on a well balanced and antibiotic free diet. The eggs were incubated at 37^0 C within a special egg incubator with 65% humidity. Eggs were turned mechanically twice a day. After 5-6 days of incubation, egg checking was done and the fertile eggs were selected by candling and the sterile eggs along with the eggs with dead embryos were rejected.

On the 12th day eggs were then candled with the help of an illuminator in a dark room to check viability, movements of embryos and define the area of blood vessels. The air space was marked on the egg shell with a pencil and a point was selected on CAM avoiding injury of large blood vessels. The air space was punctured with a pointed end of hand punch. The shell was also punctured after clearing with a sterile cotton swab on the marked spot over the CAM, using the hand punch with slight rotatory motion, avoiding injury to the shell membrane. The shell dust was blown away with capillary pipette. A drop of sterile normal saline was placed inside over the inoculation site. The tip of the blunt instrument was inserted through the drop of saline. Slight suction with a rubber teat over the hole at the blunt end of the egg was done to have a complete dropping of the membrane confirmed by candling.

The inoculum was deposited on the CAM with the help of tuberculin syringe, and the inoculated egg was rotated to facilitate the dispersion of the inoculum. The hole in the air sac was sealed and the inoculated eggs were incubated at 37^0C for 48 hrs in a horizontal position. After 48 hours the CAM was harvested aseptically after the shell membrane was broken with a blunt forceps for maximum exposure of the CAM. The membrane was cut out with a sterile pair of scissors and placed in a Petri dish for further examination.

2.6 Control study

Japanese encephalitis virus (50µl) in the same concentration (10^{-3}) in bovine albumin phosphate saline (BAPS), pH 7.20 mixed with equal volume (50µl) of sterile pyrogen free water was also inoculated on CAM. The pock count of this control virus study was considered as baseline data and any deviation from this baseline was noted after application of different medicines in the test series/group. An initial experiment was also done with different concentrations of viruses to find out the dilution which gave the maximum number of pocks on CAM (optimum dilution). If during the study, there was death of the inoculated eggs or membranes were not formed properly the data of that lot were excluded.

Apart from this virus control experiment, similar control studies were also done with the medicine without the virus in equal dilutions with sterile water. Bovine albumin in phosphate saline pH 7.2 was also studied similarly as viral dilutions were done with this buffer.

2.7 Observation of growths on CAM

Inoculated CAMs were observed after 48 hours of inoculation particularly to see the formation of pocks and other associated changes on CAM.

2.8 Standardization of the viral inoculums in suckling mice

The virulent Nakayama strain virus was consecutively passed three times in mice, and the virus suspension was prepared from the third stage which was designated as the 10^{-1} stock suspension. For determination of a 50% lethal dose (LD_{50}), several lots of suckling mice were injected intracerebrally (i.c.) with dilutions of the stock suspension from 10^{-1} to 10^{-9}. All the inoculated suckling mice were observed daily to note their survival and following this LD_{50} value was calculated by a standard method (Reed and Muench, 1938).

2.9 Methods of inoculation of suckling mice (Bandyopadhyay et al 2011)

This was done following the method described by Gould and Clegg (1985). One paper towel was placed on a cork board inside the safety cabinet. One sterile disposable syringe (1 mL) with 26-gauge sterile disposable needle was loaded with appropriate inoculums after wearing disposable gloves and with hands well inside the safety cabinet. The suckling mouse was then placed on the paper towel. It was gripped firmly and 0.02 mL of the sample was inoculated into one of the cerebral hemisphere penetrating no more than 1.5 mm. The needle is kept in position for 2 seconds then removed slowly. The procedure was repeated until the complete litter had been inoculated. For determination of LD_{50} dose, inoculations were commenced with the highest dilution, so that the same needle and syringe could be used with each viral dilution. After completing inoculations each litter was returned to its mother, the cage was properly labeled and kept in the rack. Mice showing severe disease signs or those that died within 2 h of observation were immersed in a closed vessel containing chloroform and later discarded according to statutory guidelines for Biomedical Waste management.

2.10 Method of inspection and virus collection

During inspection the condition of each mouse was recorded as D (dead), S (sick), N (normal), or M (Missing). However, in this experiment there was no M category. When the infected brain was collected for preparation of the inoculums, the dead mouse was pinned on to the cork board placed over two paper towels, one pin was placed through the nose and the other through the base of the tail. The mouse was then soaked with sufficient amount of rectified spirit and the scalp was removed using one pair of sterile scissors and forceps. This was done by cutting across the back of the scalp using scissors that have a pointed and a blunt blade. The pointed blade of the scissors was inserted into the soft rear centre point of the skull and the outside was cut down towards the nose. The skull cap was then lifted up and the brain was removed by the closed ends of the scissors.

2.11 Disposal procedure

All mice, paper towels, gloves and disposable instruments were disposed according to standard guidelines for Biomedical Waste management.

The cages and the instruments were sterilized and the safety cabinet was thoroughly cleaned with disinfectants.

2.12 Experimental design

Suckling mice (2-3 days old) were taken from litters in which mothers were orally fed with 0.06 mL of Belladonna 200 for 7/14 days. In control experiment, suckling mice were similarly taken in which mothers were not orally fed with the medicine. After this, suckling mice of both the groups were challenged with 0.02 mL of the supernatant of clarified JEV infected mice brain emulsion (10%) diluted to a LD_{50} dose intracerebrally and observed for 30 days post inoculation period. All the mice were observed daily after inoculation and every four hours after the onset of clinical signs. Clinical signs of the disease in mice particularly in last 24 hours were refusal to feeds , became disarranged in the nest, tremors and muscular spasms, ataxia, and hind-limb paralysis followed by death within a few hours. Those suckling mice that died within the first 24 hours was considered as non-specific deaths.

For preparation of infected brain emulsion required for LD_{50} determination and standardization of the inoculums, brains were collected close to the time of death. When the animal showed acute signs of uniform sickness (usually on third post inoculation day), their brains were harvested aseptically. The brains after weighing, were ground in a homogenizer (Lourde's homogenizer) placed in ice bath to give a 10% w/v suspension in BAPS with antibiotics. The suspension was initially centrifuged at 2000 rpm at 4 degree C for 15 minutes, following which the supernatant was recentrifuged at 10,000 rpm at 4 degree C for one hour (in a refrigerated centrifuge). The supernatant was carefully collected and was kept in small aliquots of 0.5 ml screw capped vials, labeled and sealed and stored at - 70 degree C deep freezer after shell freezing. Throughout the entire study, the passage level of the stock mice was maintained constant. For this purpose, stock of the virus, one or two passages prior to the working stock was stored at - 70 degree C. Whenever the working stock got exhausted or the titer of the stock fell short by one log due to prolonged storage suckling mice were inoculated with the back passage material and fresh working stock of the same passage level was prepared.

2.13 Histological studies

Macroscopically presence of congestion, swelling, petechial haemorrhages and microscopically evidences of neuronophagia, microglial nodules, intranuclear/intracytoplasmic inclusion bodies, inflammatory cell cuffing in the Virchow-Robin spaces, reactive features of astrocytosis, diffuse microglial activation were studied with the brain specimens taken out from the suckling mice in different experimental groups including the control group.

3. Results

3.1 CAM experiment

Optimum dilution of the viruses was studied with different concentrations of the viruses ~ Neat (10%), 10^{-1}, 10^{-2},10^{-3},10^{-4},10^{-5}, control (buffer solution without viruses). The results showed that 10^{-3} dilution showed maximum pock count and thus this optimum concentration was used throughout the experiment. Results of different experiments with

Belladonna 200 are given in Table 1. Control studies with different medicines and bovine albumin phosphate saline showed normal findings.

Experiment	Pock count on CAM in number (Average±SD±SEM)	t-value of the difference and its significance
Belladonna 200 (N-300)		
Virus control	53.97±28.21±4.70	6.95, P value highly significant at 0.01 level
Virus + Medicine	18.17±12.66±2.11	

CAM= Chorioallantoic membrane, SD=Standard deviation, SEM=Standard error of mean. N=Number of inoculated eggs. Eggs that were dead or yielded deformed or absent CAM, were not considered for calculation of the results.

Table 1. Changes in pock count on CAM with *"Belladonna 200"*.

3.2 Suckling mice experiment

LD_{50} of the virulent Nakayama strain JE virus was 7.0 log_{10} /0.02mL. The average survival rate of the control group (n=96) where mothers were not treated with the medicine was 47.92% (Fig.1), similar rate in the experimental group (n=67) of suckling mice where mothers were treated with Belladonna 200 for 7 d was 80.60% (Fig.2), and the rate in the last experimental group (n=53) of suckling mice where mothers were treated with the medicine for 14 days was 79.24%. Details of the outcome of this study and statistical analysis (χ^2 test) are given in Table 2. The statistical analysis (χ^2 test) indicated a highly significant difference with p value significant at 0.01 levels. Histological studies showed variable changes in the virus inoculated control group while in the test group there was no change in any of the survived mice after the challenge.

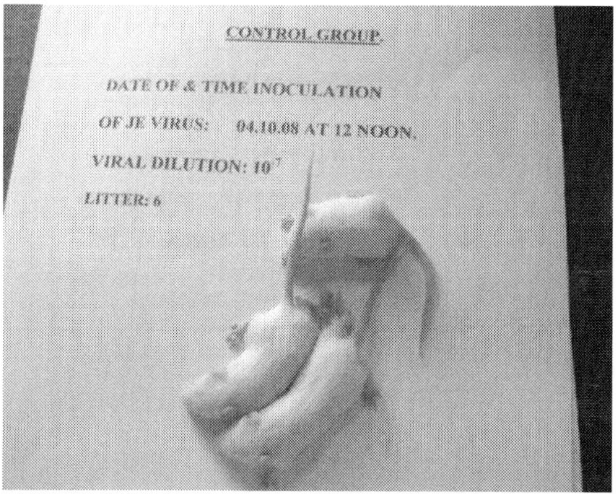

Fig. 1. Three alive mice in control group (3 survived out of 6 inoculated suckling mice, lot 1): 30 days observation

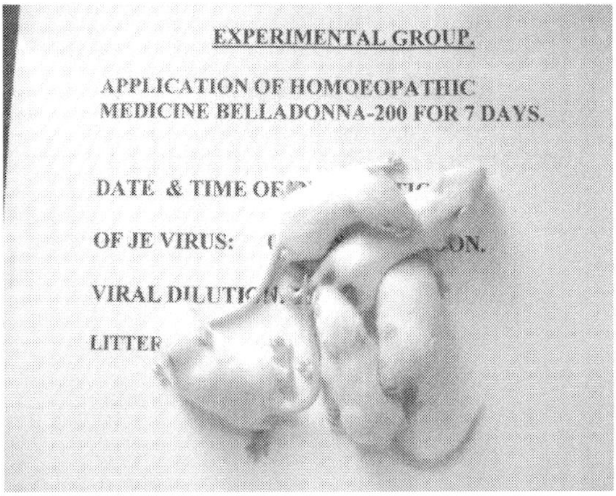

Fig. 2. Five alive mice in Experimental group1 (5 survived out of 6 inoculated suckling mice, lot 1): 30 days observation

Parameters	Control group: (Sucklings of untreated mother mice)	Experimental group 1: (Sucklings of Mother mice treated with Belladonna 200 for 7 days)	Experimental group 2: (Sucklings of Mother mice treated with Belladonna 200 for 14 days)	Total
Attacked with JE	50	13	11	74
Not attacked with JE	46	54	42	142
Total	96	67	53	216
Expected attacks (50%)	48	33.5	26.5	
(Observed-Expected)^2	4	420.25	240.25	
(Observed-Expected)^2 /(Expected) (χ^2 test)	0.08	12.54	9.07	21.69(p value significant at 0.01 level)

Table 2. Statistical analysis (χ^2 test) showing significant protective action of Belladonna 200 in suckling mice challenged with LD_{50} dose of virulent Nakayama strain of JE virus when mothers were fed with the medicine.

4. Discussion

Japanese encephalitis(JE) is basically a zoonotic disease where ardeoid water birds mainly herons and egrets are the reservoirs with frequent spread out even in epidemics to pigs, members of the family of equidae mainly horses and donkeys as well as in humans. The central vector of the disease is zoophilic *Culex tritaeniorhynchus* related to the *Cx. gelidus* complex. It principally affects central nervous system and can produce severe neurological complications and even death (Okuno, 1978).

This disease is currently prevalent in south Asia, south east Asia, east Asia and in the Pacific with a 3 billion population at risk, an annual incidence of 30,000 to 50,000 cases, with 10,000 to 15,000 annual deaths (Solomon, 2006) and a global impact of about 7,00,000 disability-adjusted life years (DALYs)(WHO,2008). Local incidence rates usually range from 1-10 cases per 100,000 populations but can reach more than 100 cases per 100,000 populations during outbreaks. Among countries particularly India, Nepal, Thailand, Malaysia, Myanmar, Japan, China, Indonesia, Bangladesh extending in a wide zone from Pakistan to Siberia and Japan where the disease is now prevalent, in seven countries JE virus infection is presently increasing with a significant percentage of total population living in rural JE-endemic area (Fig.3) and among these countries DALYs are the greatest in India (Fig.4).

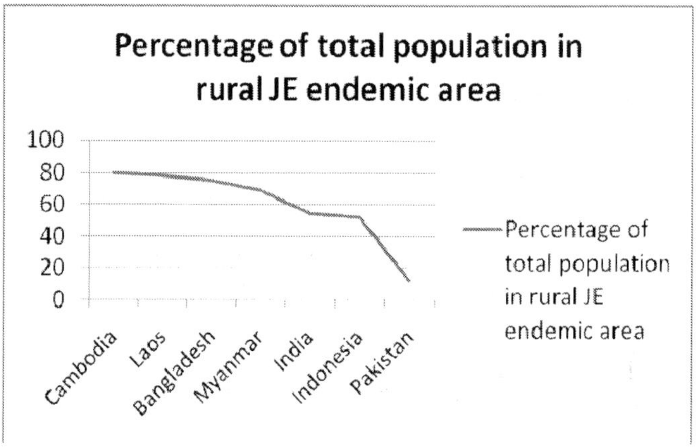

Fig. 3. Percentage of total population living in rural JE endemic area in countries where JE virus infection is increasing.

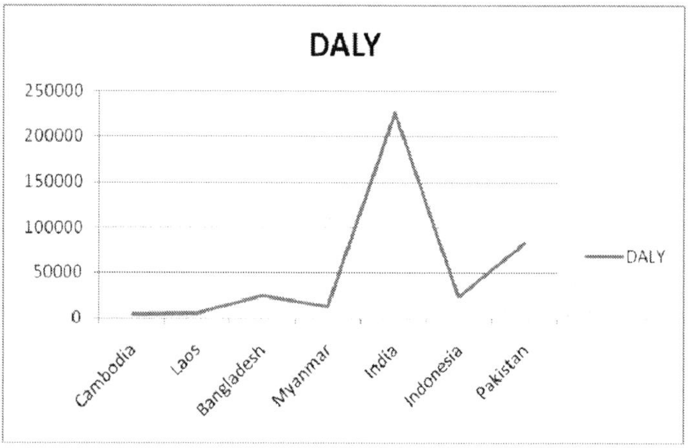

Fig. 4. Disability-adjusted life years (DALYs) in countries where JE virus infection is increasing (prepared with data obtained from various sources).

Recently more than 1300 children died in outbreaks of JE in the north eastern state of Uttar Pradesh, although possibility of other enterovirus infection mixed with JE virus has not been ruled out.

Live attenuated vaccine against JE was developed about 40 years back (Igarashi,2002) and newly developed vaccines like live inactivated yellow fever virus - chimeric vaccine (Solomon, 2006) and adjuvanted SA 14-14-2 vaccine (E. Tauber and S. Dewasthaly, 2008) grown on Vero cell lines, are not only safe but also effective in single doses. However, two important obstacles for an effective vaccination program are there which are very difficult to solve: difficulties in delivering the vaccine in rural poor population and failures of vaccination in newborn pigs having maternal antibodies and with a high turnover rate of their population.

Experiments on intermittent irrigation (alternate wetting and drying) of paddy fields was found effective in vector control but its implementation is also very complicated because all paddy fields could not be covered simultaneously in such a program due to a vast area, sufficient water may not be available in rice-growing season and countrywide educational as well as supportive program may not be practicable (Rajendran et al, 1995). The transmission of JE virus is multi-factorial with at least five variables directly or indirectly influences the rate at which the virus is transmitted - the viral strain, vector, wild vertebrate hosts, humans, and environmental factors; thus incomplete targeting to one factor has got very little influence in this complex system of transmission.

Alterations in temperature and rainfall patterns induced by recent global climatic changes may lead to a significantly increased vector population in the endemic areas which will be very difficult to control.

Chemical control of the vectors with pyrethroids, organophosphates, carbamates etc. was also not found suitable due to their short term effects and rising levels of insecticide resistance. (S. H. Karunaratne and J. Hemingway, 2000).

The disease usually starts as a flu-like illness with fever, headache, nausea, vomiting, weakness. Altered mental status usually occurs from mild confusion to agitation to coma. Seizures develop in about 66% children, while headache and meningismus are commonly found in adults. There is no specific medicine to treat the patients suffering from JE. According to the World Health Organization the disease is fatal in up to 30 percent of cases, and there is a possibility that those who survive may be disabled for life.

The homeopathic medicine Belladonna 200 is prepared with the root, and the leaves of *Atropa belladonna*, which is also known as Deadly Nightshade, Dwale, Black Cherry, Strygium and Strychnon. In ancient times, the Venetians named this plant as Belladonna because at that time ladies used a distilled product of the plant as a cosmetic; hence the name "Bella-donna" or beautiful lady. One important indication of Belladonna is its use as a medicine in the treatment of patients with cerebral congestion as mentioned in homoeopathic pharmacopoeia. In a previous study we also observed preventive role of Belladonna in JE virus infected chick chorio allantoic membrane (Bandyopadhyay et al, 2010).

Mosquitoes in the genus *Culex* have a strong tendency to ornithophagy probably due to relatively inactive thrombocytes present in birds because anti-thrombin peptides are probably less in the saliva of these mosquitoes, only recently they are gradually adapted to mammals. It appears that recent changes due to global warming will accentuate their behavior in adaptation and a great havoc in human population may occur in near future by this virus unless we are prepared.

Medicinal properties of Belladonna are known from times immemorial. Many homeopathic practitioners used it in prevention of JE, although there is no experimental proof in favor of it. This experiment clearly indicated preventive role of this medicine against JE which is statistically highly significant.

Although such action of Belladonna is difficult to explain we here propose a hypothetic action pathway of this medicine in preventing JE virus infection.

If we look into the occurrence of calystegines and related compounds in *A. belladonna* and related plants then it is obvious that these are present in significantly higher amounts (Fig.5) in *A. belladonna* (Draeger et al, 1995).

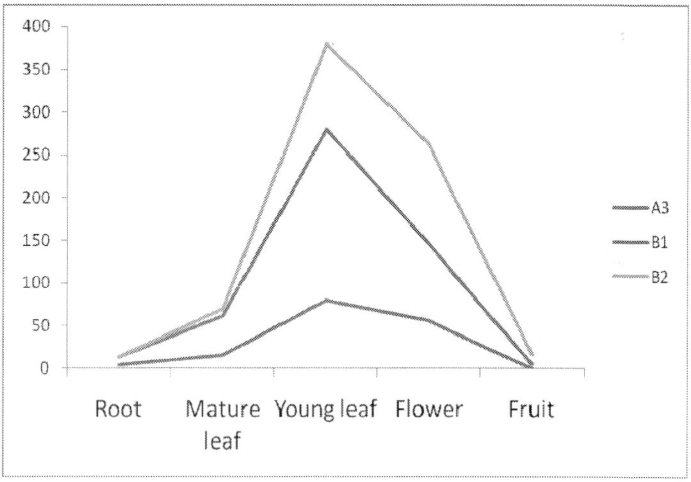

Fig. 5. High levels (µg/g fresh mass) of Calystegines (A3, B1, B2) in *Atropa belladonna* (based on the report by Draeger et al, 1995)

Calystegines are well known selective glycosidase inhibitors in comparison to common tropane alkaloids atropine and scopolamine of *A. belladonna*, which are parasympatholytic. Like most glycosidase inhibitors, calystegines compete with the substrate for binding to the active site as observed in kinetic interaction measurements. Most glycosidases perform enzymatic hydrolysis reaction with the aid of a glutamic acid residue in the active cleft and calystegines mimic the transition state of this reaction.

Most enveloped viruses like human immunodeficiency virus, hepatitis B virus etc. showed altered life cycle during invasion of cells in which glucosidase-mediated N-linked oligosaccharide trimming is inhibited and N-linked oligosaccharide processing events in the endoplasmic reticulum are important for the secretion of some enveloped viruses (Mehta et al, 1998) characterized by sequential trimming of the glucose residues on oligosaccharide precursor.

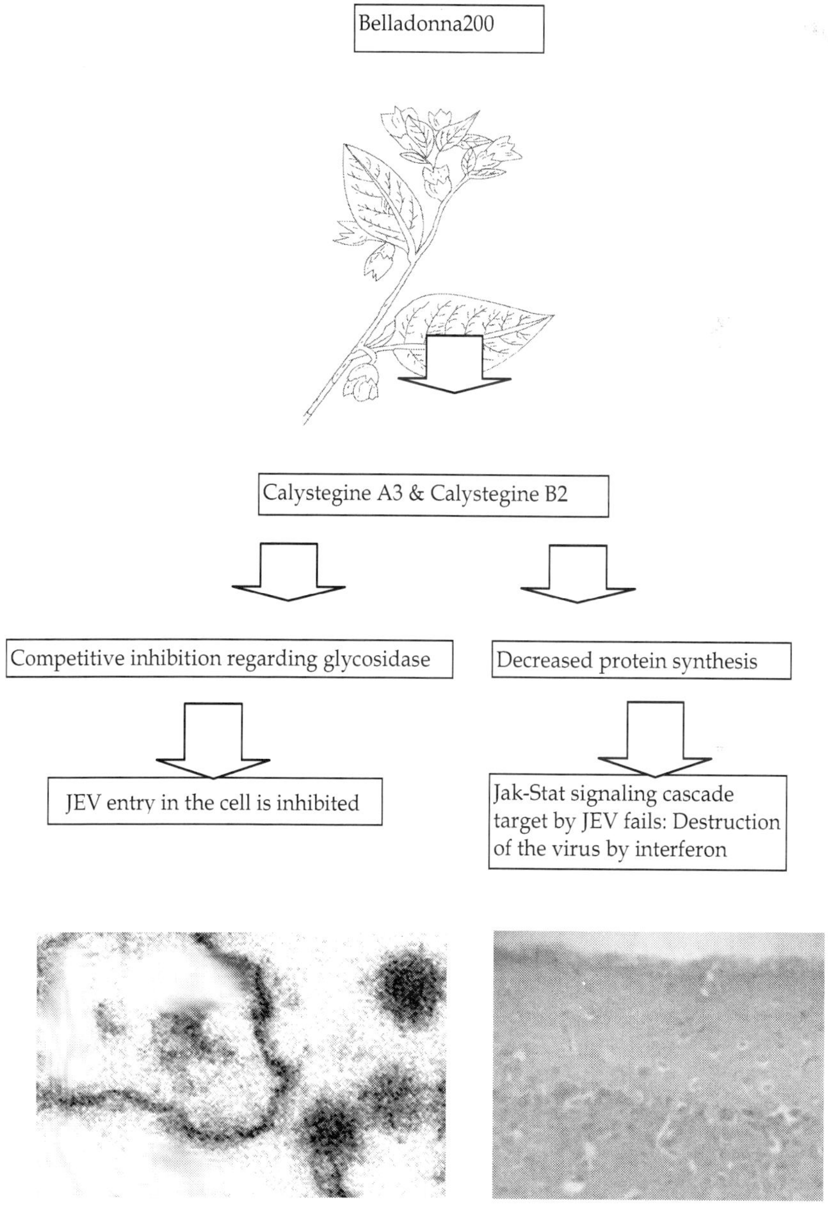

Fig. 6. The possible pathway of the action of Belladonna200 in prevention of JEV infection in mice.

It has also been demonstrated that Dengue virus envelope glycoprotein processing in cells was strongly affected by this important mechanism (Courageot et al, 2000). Thus Belladonna may appear to act through this pathway in preventing JE virus infection in suckling mice (Fig.6). This also may lead to inhibition of the synthesis of key amino acid residues of the E protein of Nakayama strain JE virus(E107,Leu; E138,Glu; E176,Ile; E177,Thr; E264,Gln; E279,Lys; E315,Ile; E439,Lys) which are important for neurovirulence. It also blocks JE virus in evading action of interferon on them. Although interferon has been identified as the most promising antiviral agent against JE virus (Crance et al, 2003), but inside the body JE virus usually target the Jak-Stat (Janus Kinase-signal transducer and activation of transcription) signaling cascade to evade the interferon response leading to failure of interferon treatment in JE infected children (Solomon et al, 2003). In this connection, it is important to note that cluster of genes encoding the interferon - induced 2′-5′-OAS (2′,5′-Oligoadenylate synthetases) is present in chromosome 5 in mice- known also as Flv gene is responsible for susceptibility to JE virus infection in mice. (Perelygin et al, 2002). It has been observed that some newly synthesized proteins are required to block Jak-Stat signaling by JE virus (Lin et al, 2004). Thus protein synthesis inhibitors may play a role in decreasing pathogenecity of JE virus where this medicine may play a role through the above mentioned pathways.

5. Conclusion

In a search to find out a newer method to prevent JE infection, we have studied *"Belladonna 200"* - a homeopathic medicine as a possible candidate which may fulfill our aim. Experiments utilizing two different models- CAM model and suckling mice model, both conclusively indicated that this medicine has got a definite role in preventing JE, the mechanism of which is still unknown. During the experiment we also standardized these two models which may help all future workers in this line.

6. Acknowledgements

This study was done as a part of the project study (No.17-88/2006-2007/CCRH/Tech/Coll/STM) financed by CCRH, AYUSH, Govt. of India. We thank Prof P.N.Gupta, Ex-Head of the Department of Microbiology, Medical College and Hospital, Kolkata for his valuable suggestions. We are thankful to the Institutional authorities and all the faculties of the Microbiology Department of School of Tropical Medicine, Kolkata, for their continued support and inspiration. We are also thankful to our technical personnel Surja Kumar Halder, Suniti Bikash Ghosh, Astom Chandra Adak, Pradip Ray, Kajori Mukherjee and Kartick Chandra Roy, for their dedicated participation in the project.

7. References

Bandyopadhyay B, Das S, Sengupta M, Saha C, Das K C, Sarkar D, Nayak C. 2010, Decreased intensity of Japanese encephalitis virus infection in chick chorioallantoic membrane under influence of ultradiluted belladonna extract. American J. Infectious Diseases, 6(2):24-28. http://www.scipub.org/scipub/patch.php_

Bandyopadhyay B, Das S, Sengupta M, Saha C, Bhattacharya N, Raveendar C, Chakravarthy R, Ray K, Nayak C. 2011, Suckling Mice of "Belladonna 200" Fed Mothers Evade Virulent Nakayama Strain Japanese Encephalitis Virus Infection. International Journal of Microbiological Research, 2 (3): 252-257.

Courageot M, Frenkiel M, Santos C D D, Deubel V, Despres P. 2000, α-Glucosidase inhibitors reduce Dengue virus production by affecting the initial steps of virion morphogenesis in the endoplasmic reticulum, Journal of Virology, 74:1, 564-572. DOI:10.1128/JVI.79.14.8698-8706.2005

Crance JM, Scaramozzlno N, Jouan A, Garin D. 2003,Interferon, ribavirin, 6-azauridine and glycyrrhizin: antiviral compounds active against pathogenic flaviviruses. Antivir. Res. 58:73-79. DOI:10.1016/S0166-3542(02)00185-7

Draeger B, Van-Almssick A, Mrachatz G. 1995, Distribution of calystegines in several Solanaceae. Planta Medica, 61, 577-579. DOI:10.1093/jac/dkm 503

Igarashi A. 2002, Control of Japanese encephalitis in Japan: immunization of humans and animals, and vector control. Curr Top Microbiol Immunol, 267:139-52. DOI:10.3201/eid1411.071368

Karunaratne S H, Hemingway J. 2000, Insecticide resistance spectra and resistance mechanisms in mosquitoes, *Culex tritaeniorrhynchus* and *Cx. gelidus*, in Sri Lanka. Med Vet Entomol, 14:430-6. www. the free library.com.

Lin R-J, Liao C-L, Lin E, Lin Y-L. 2004, Blocking of the alpha interferon -induced Jak-Stat signaling pathway by Japanese encephalitis virus infection, Journal of Virology, 78:17, 9285-9294. DOI:10.1128/JVI..78.17.9285-9294.2004

Gould E A, Clegg J C S. 1985, Growth, titration and purification of Togaviruses, Mahy B W J(Ed) In Virology - a practical approach, IRL Press, Oxford, pp 43-74.

Mehta A, Zitzmann N, Rudd P M, Block T M, Dwek R A. 1998, α-Glucosidase inhibitors as potential broad based anti-viral agents, FEBS Lett. 430: 17-22.

Okuno T. 2006, An epidemiological review of Japanese encephalitis. World Health Stat Q. 1978, 31:120-33. Indian J Med Res 124, pp 211-212.

Perelygin AA, Scherbik SV, Zhulin IB, Stockman BM, Li Y, Brinton MA. 2002, Positional cloning of the murine flavivirus resistance gene. , Proc. Natl. Acad. Sci. USA, 99:9322-9327. DOI:10.1073/pnas.142287799.

Rajendran R, Reuben R, Purushothaman S, Veerapatran R. 1995, Prospects and problems of intermittent irrigation for control of vector breeding in rice fields in southern India. Ann Trop Med Parasitol, 89:541-9. PMID: 7495368

Raue C G. 1885, Special pathology and diagnostics with therapeutic hints, 4th Ed, reprint 1975, Sett Dey & Co, Calcutta, pp 915.

Reed L J, Muench H. 1938, A simple method of estimating fifty percent endpoints. Am J Hyg, 27:493-7.

Solomon T, Dung NM, Wills B, Kneen R, Gainsborough M, Diet TV, Thuy TT, Loan HT, Khanh VC, Vaughn DW, White NJ, Farrar JJ. 2003, Interferon alfa-2a in Japanese encephalitis: a randomized double-blind placebo-controlled trial. Lancet, 361: 821-826. DOI: 10.1016/S0140-6736(03)12709-2. 38.

Solomon T. 2006, Control of Japanese encephalitis - within our Grasp? N Engl J Med, 355:869-71. DOI:10.1056/NEJMp058263 PMID:16943399.

Tauber E, Dewasthaly S. 2008, Japanese encephalitis vaccine - needs, flaws and achievements, Biol Chem, 389:547-50.

www.who.int/vaccine_research/diseases/vector/en/index5.html

World Health Organization. World Health Report (for years 2000-2004) 2008, available from http://www.who.int/whr/en.

8

Interactions Between Nucleopolyhedroviruses and Polydnaviruses in Larval Lepidoptera

Vincent D'Amico[1] and James Slavicek[2]
[1]*USDA Forest Service, University of Delaware, Newark, DE,*
[2]*Delaware, OH,*
USA

1. Introduction

The field dynamics of some insect populations are strongly influenced by two types of insect viruses: the nucleopolyhedroviruses (NPVs) and the polydnaviruses (PDVs). Although greatly different in origin and mode of infection, both viruses produce considerable mortality directly and indirectly in the field, and have evolved reproductive strategies that use the same life stage of the same host insects. The life histories of these host insects are such that there are many opportunities for coinfection and competition between baculoviruses and polydnaviruses, although these relationships remain largely theoretical and unexplored.

The hosts of both these viruses are found primarily in the insect order Lepidoptera, the moths and butterflies. The viruses are members of two large and diverse families: the Baculoviridae, pathogens of insects that fit the common conception of disease-causing organisms, and the Polydnaviridae, genome-integrated wasp symbionts with a unique natural history. Although the Baculoviridae are known from several insect orders, they occur in their greatest variety as the nucleopolyhedroviruses in the larval Lepidoptera. This is probably because the relatively long, plant-feeding caterpillar stages of moths and butterflies give ample opportunities for *per os* infection, as we will describe in detail below. Many nucleopolyhedroviruses known from the Hymenoptera typically occur in families feeding on leaves during the larval stage, as well, but they will not be included here. The ease of rearing lepidopteran larvae, and their place as pests of human crops, has also led to an accumulation of data on their viral diseases, and the many nucleopolyhedroviruses infecting Lepidoptera make this insect order a good starting point for exploring interactions with other viruses known to occur within them, polydnaviruses in particular. As symbionts of parasitoid wasps, polydnaviruses are also found in lepidopteran larvae. There they produce products that abrogate the larval immune system. Larval hosts immunosuppressed by a polydnavirus fail to muster the encapsulation response that would otherwise kill the parasitoid eggs and larvae; the wasp mutualist, or *carrier*, brings the polydnavirus to its host to ensure the survival of its own progeny. Our interest comes from past research in the field on population and disease dynamics of outbreaking lepidopteran forest pests, although we do not limit ourselves to these here.

2. Nucleopolyhedroviruses

Baculoviruses are a large group of viruses pathogenic to arthropods, primarily insects from the order Lepidoptera and also insects in the orders Hymenoptera and Diptera (Moscardi 1999; Herniou & Jehle, 2007). These viruses have been isolated from over 300 insect species (David, 1975; Tinsley & Harrap, 1978; Harrap & Payne, 1979). Baculoviruses have been used to control insect pests on agricultural crops and forests around the world (Moscardi, 1999; Szewczk et al., 2006, 2009; Erlandson 2008). The *Baculoviridae* are divided into four genera: the Alphabaculovirus (lepidopteran-specific nucleopolyhedroviruses, NPV), Betabaculovirus (lepidopteran specific Granuloviruses, GV), Gammabaculovirus (hymenopteran-specific NPV), and Deltabaculovirus (dipteran-specific NPV) (Jehle et al., 2006). Baculoviruses are arthropod-specific viruses with rod-shaped nucleocapsids ranging in size from 30-60 nm x 250-300 nm.

NPVs initiate infection when a susceptible host ingests virus in the form of occlusion bodies (OBs) present on host plants (Fig. 1A). OBs are composed primarily of the protein polyhedrin, which forms a paracrystalline matrix into which occlusion derived virus (ODV) is embedded. The polyhedron provides the embedded ODV protection from environmental elements such UV light. NPVs produce another form of virus, termed the budded virus, that is produced in the early stages of viral replication (Rohrmann, 2011). Within the alkaline environment of the larval midgut, OBs dissolve, thereby releasing ODV. To initiate an infection, ODV must first traverse a physical structure termed a peritrophic membrane (Fig. 1B), which is composed of proteins, mucopolysaccharides, and chitin (Pritchett et al., 1984; Hegedus, et al., 2009). The peritrophic membrane provides a barrier to gut cells to bacteria, viruses, fungi, and physical damage from ingested plant material. The peritrophic membrane is in a constant state of regeneration from epithelial cells as larvae feed, and the movement of food material through the insect gut also causes loss of the peritrophic membrane.

After penetration of the peritrophic membrane, ODV gains entry into midgut cells by a type of fusion process (Fig. 1C), although this has resisted definitive characterization (Granados & Lawler, 1981). The type NPV, *Autographa californica* multiple NPV (AcNPV), initiates the infection cycle by infecting columnar epithelial cells within the midgut and regenerative epithelial cells in *Trichoplusia ni* (Keddie et al., 1989) or *Spodoptera exigua* larvae (Flipsen et al., 1995). Several factors are involved with the initial act of infection that includes ODV binding to midgut cells at cell receptors, and viral entry into the cells. All sequenced baculoviruses contain genes that code for *per os* infectivity factors (PIFs) that are associated with ODVs but not budded virus (Faulkner et al., 1997; Kikhno et al., 2002; Fang et al., 2006; Harrison et al., 2010; Fang et al., 2009). The pif genes include p74-pif, and pif genes 1-5, Ac119, Ac22, Ac115, Ac96, and Ac148, respectively. Deletion of any of the genes from a viral genome significantly decreases but does not eliminate *per os* infectivity (d'Alencon et al., 2004; Crouch et al., 2007). The PIFs, with the exception of PIF3, are thought to be involved in binding or interacting with the midgut cells that leads to infection (Ohkawa, et al., 2005; Li et al., 2007; Peng et al., 2010; Horton & Burand, 1993).

Upon entry into midgut cells the nucleocapsids are actively transported to nuclear pores in a process that uses actin polymerization (Ohkawa et al., 2010) (Fig. 1C). Viral DNA is then released into the cell nucleus and viral replication ensues (Rohrmann, 2011) (Fig. 1C).

During the early phase of viral replication, BV are produced that bud from midgut cells and infect tracheal epidermal cells, which penetrate the basal lamina (Volkman, 2007). Infection spreads via the tracheal system and haemocytes until many different cells are infected (Engelhard et al., 1994) (Fig. 1D). During the later phase of viral replication, ODV are produced and packaged within OBs. Upon the host's death, liquefaction occurs, releasing OBs into the environment to infect another host (Reardon, 1996; Riegel & Slavicek, 1997).

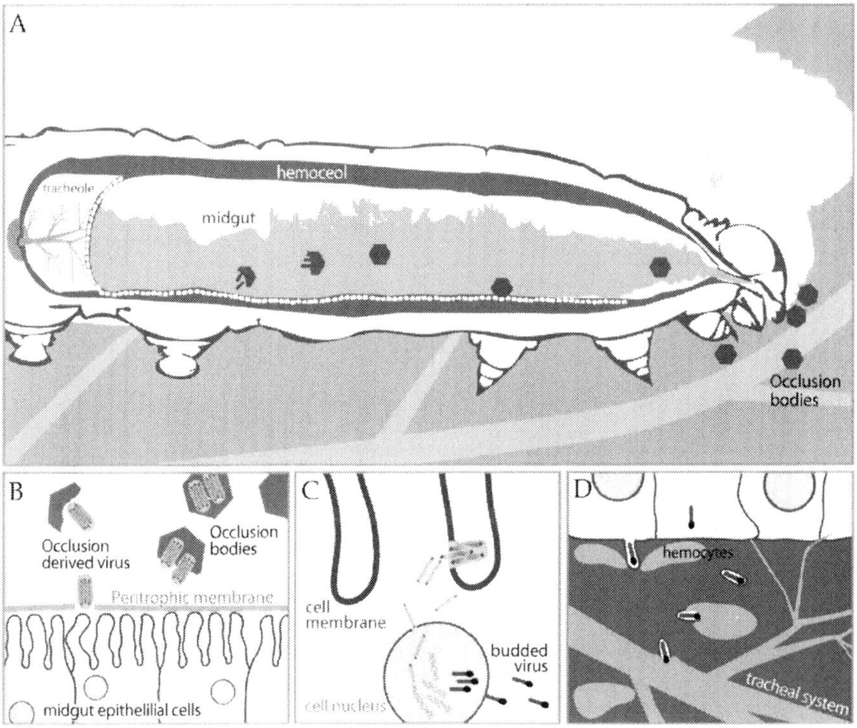

Fig. 1. Nucleopolyhedrovirus transmission. (A) NPV infections in larval Lepidoptera are initiated when caterpillars ingest viral occlusion bodies on foliage. (B) These dissolve in the alkaline conditions of the midgut and release occlusion derived nucleocapsids, which penetrate the peritrophic membrane. (C) Nucleocapsids fuse with the membrane of epithelial cells and pass through nuclear pores to begin production of the budded form of NPV. (D) Budded virus leave the midgud cells and infects hemocytes and other cell types.

Insect hosts have developed several mechanisms to thwart baculovirus infections including the peritrophic membrane physical barrier, developmental resistance (Engelhard & Volkman, 1995), melanization of infected cells, and apoptosis of infected cells. The *Lymantria dispar* MNPV (LdMNPV) exhibits a LD50 (74 OBs) in *L. dispar* (Hoover et al., 2002); however, by the middle of the fourth instar, the LD50 of LdMNPV is 18-fold higher than in newly molted larvae. The majority of cases of developmental resistance within an instar are midgut-based (Haas-Stapleton et al., 2003). As the virus infects midgut cells the cells are in a continual process of renewal and sloughing. Consequently, if BV are produced and escape the midgut cell before it sloughs off and is excreted, the virus can infect the tracheal cells and usually generate a systemic infection (Engelhard et al., 1994). However, loss of tracheal epidermal cell infections can occur (Haas-Stapleton et al., 2003), and consequently the host can escape viral infection. In addition, larval hosts are less susceptible to viral infection between and within larval instars, due to a process termed developmental resistance (Engelhard & Volkman, 1995).

Because the main way in which polydnaviruses might affect NPV pathogenesis is via immunosuppression, it is important to note that the immune response is a vital part of larval response to infection. Hoover et al., (2002) hypothesized that immune responses play a role in systemic resistance in *L. dispar* to LdMNPV. In non- or semi-susceptible hosts, cellular immune responses to AcNPV-infected tissues can occur (Rivkin et al., 2006; Trudeau et al., 2001; Washburn et al., 2000). Haemocytes of some hosts are refractory to AcNPV infection and replication, which slows or stops the spread of the virus within the haemocoel (Rivkin et al., 2006). Apoptosis of AcNPV-infected cells can also serve as an antiviral response (Clarke & Clem, 2003; da Silveira et al., 2005; Zhang et al., 2002). In a recent study using LdMNPV, McNeil et al. (2010a) found that encapsulation and apoptosis of infected tissues are mechanisms of host defense. Elimination of infected tissue through encapsulation and melanization can occur in lepidopterans infected with AcNPV, but has only been reported in non-susceptible hosts such as *Helicoverpa zea* and *Manduca sexta* (Trudeau et al., 2001; Washburn et al., 2000) and did not differ in effectiveness at different times within the instar. Apoptosis of tracheal epidermal cells was hypothesized to contribute to AcNPV resistance in *Spodoptera frugiperda* (Haas-Stapleton et al., 2003) and it appears to be an effective defense against AcNPV infection in the fat body and epithelium of IH-injected *S. frugiperda* (Clarke & Clem, 2003) and *S. exigua* (Clem, 2005). Removal of midgut infections in *L. dispar* probably involves apoptosis of infected midgut epithelial cells (Dougherty et al., 2006) and is considered a conserved mechanism for midgut based resistance to NPVs in most lepidopterans (Volkman, 2007).

A recent study by McNeil et al. (2010b) found that mid-instar *L. dispar* larvae exhibited a higher degree of hemocyte immunoresponsiveness, a greater potential hemolymph phenoloxidase (PO) activity at the time the virus is escaping the midgut to enter the hemocoel, greater FAD-glucose dehydrogenase (GLD) activity, and more targeted melanization of infected tissue, which correlate with reduced viral success in the host. PO and GLD are components of the humoral immune response associated with melanized encapsulation (Lovallo & Cox-Foster, 1999; Nappi & Christensen, 2005). Phenoloxidase has potent anti-microbial properties, and its activation is tightly controlled within the hemolymph to prevent non-specific activation from harming the host (Jiravanichpaisal et al., 2006). PO activity in the hemolymph of *Heliothis virescens* larvae has been shown to have virucidal effects in vitro (Shelby & Popham, 2006) and previous research has supported the role of PO and cellular immunity in immune defenses against viruses (Stanley & Shapiro, 2007, 2009; Shrestha & Kim, 2008). GLD is induced during encapsulation responses to pathogens (Cox-Foster & Stehr, 1994; Lee et al., 2005). GLD is also involved in the production of free radicals derived from quinones during melanization in insects, and has been hypothesized to strengthen melanized capsules (Cox-Foster & Stehr, 1994), which could also negatively impact viral success.

3. Polydnaviruses

Polydnaviruses are multipartite double-stranded DNA viruses originating in koinobiont endoparasitic Hymenoptera, commonly called parasitoid wasps. The polydnavirus genome is integrated into that of the "carrier" wasp species, and is transmitted through the germ line (Stoltz, 1990; Belle et al., 2002; Wyler & Lanzrein, 2003; Bezier et al., 2009). The origin, replication, and function of polydnaviruses, and their symbiotic relationship with their carriers, make this system unique among members of any taxon. What follows is a short review of the discovery of the polydnaviruses, and our current understanding of their reproduction and functions. We will use the term *carrier* to refer to wasps bearing an genome-integrated polydnavirus. This avoids the connotations of *host* and possible confusion with the host of the wasp - here a lepidopteran larva, or caterpillar. Similarly, *polydnavirus infection* is used to refer to the polydnavirus in the larval host of the carrier wasp, not in the wasp itself.

The discovery of polydnaviruses began with the work of George Salt. Salt had a career that spanned more than a half-century, beginning in the 1920s. His research on the underlying mechanisms of insect immunity culminated in the treatise, *The Cellular Defence Reactions of Insects* (Salt, 1970). This synthesis included the results of earlier work, the most pertinent of which was an experiment involving placing washed eggs of the parasitoid *Venturia (Nemeritis) canescens* into living larvae of its host *Ephestia kuehniella* (Salt, 1965; see also Rotheram, 1973). Washed eggs were encapsulated by hemocytes within the larvae, while intact unwashed eggs were not. Salt used simple light microscopy to find the origin of this resistance; a "coating" on the eggs which originated in the calyx of the female parasitoid. This work certainly informed the subsequent seminal research of Stoltz, Vinson, and Mackinnon (Stoltz et al., 1976; Stoltz & Vinson, 1977, 1979). Surveys and research led by these and other researchers in the 1970s and 1980s led to a preliminary understanding of the origin and function of the baculovirus-like particles seen in the calyx fluid of some ichneumonids and braconids (Stoltz et al., 1981; Cook & Stoltz 1983; Fleming & Summers 1986). These particles were injected by carrier wasps into larval hosts along with eggs and toxins, and expressed genes while in that host, but did not replicate in host cells. Parasitoid eggs from such wasps, when placed in a host without other calyx-derived products such as the virus-like particles and wasp toxin, were invariably recognized as foreign, encapsulated, and destroyed (Edson et al., 1981; Stoltz & Guzo, 1986). The particles were clearly necessary for successful reproduction of the carrier wasp, but the emergent question was how these baculovirus-like particles were acquired or produced. The advent of modern molecular analyses allowed researchers to answer this question definitively.

The baculovirus-like particles discovered in braconid and ichneumonid wasps were determined to be a new form of virus, integrated into the genome of the carrier wasp as proviral genetic sequences (Stoltz, 1990; Bezier et al., 2009). Although these sequences were found in both male and female wasp genomes, they were excised to form free viral particles only in the ovary calyx cells of maturing female wasps. This is in stark contrast to a typical baculovirus life cycle (Cory & Myers, 2003). In the polydnavirus life cycle, replication and vertical transmission occur in the carrier wasp, while a form of horizontal transmission and deleterious infection occurs in the carrier wasp's larval host. This has also been previously defined as the two "arms" of the PDV life cycle (Turnbull & Webb, 2002). The larval host takes the brunt of the effects of the polydnavirus in the form of immunosuppression caused by polydnavirus DNA expression - but not through polydnavirus replication in and subsequent destruction of its cells.

Most of the known polydnaviruses are carried by wasps that parasitize larval Lepidoptera, in subfamilies of the Ichneumonidae (these polydnaviruses are termed ichnoviruses) and Braconidae (the bracoviruses) (Stoltz & Vinson, 1977; Fleming & Summers, 1991; Stoltz & Whitfield, 1992; Webb & Strand, 2005). At this time, known polydnavirus carrier wasps occur only in the braconid subfamilies Cardiochilinae, Cheloninae, Mendesellinae, Khoikhoiinae, Miricinae and Microgastrinae, and ichneumonid subfamilies Campopleginae and Banchinae (Table 1) (Stoltz et al., 1981; LaPointe et al., 2007; Whitfield & O'Connor, 2012). Even if only currently named species are considered, these subfamilies contain a total of almost 30,000 wasp species, with many expected to harbor polydnaviruses. Although these braconid and ichneumonid PDVs are at least as diverse as the wasp families themselves and are apparently unrelated to each other, we may use the braconid lineage as an example of how this association originated (Whitfield, 1997; see the excellent review by Whitfield & O'Connor, 2012). Between 85-100 million years ago, an early braconid acquired an integrated nudivirus in a manner not fully understood (Banks & Whitfield, 2006; Murphy et al., 2008). This virus has evolved along with the carrier wasps to the present day, with greater or lesser diversity as a function of the diversification of the carrier wasp (Murphy et al., 2008; Bezier et al., 2009). Modern inexpensive sequencing techniques continue to provide ever-greater detail of these phylogenies.

Polydnavirus	**Wasp family**	**Subfamilies**	**Representative wasp genera**
Bracovirus (BV)	Braconidae	Cardiochilinae	*Cardiochiles, Toxoneuron*
		Cheloninae	*Adelius, Chelonus, Ascogaster, Phanerotoma*
		Khoikhoiinae	*Sania, Khoikhoia*
		Mendesellinae	*Epsilogaster, Mendesella*
		Microgastrinae	*Apanteles, Cotesia, Diolcogaster, Dolichogenidea, Glyptapanteles, Hypomicrogaster, Microgaster, Microplitis, Pholetesor*
		Miracinae	*Mirax*
Ichnovirus (IV)	Ichneumonidae	Campopleginae	*Campoletis, Campoplex, Dusona, Hyposoter, Sinophorus, Venturia*
Banchovirus	Ichneumonidae	Banchinae	*Banchus, Glypta*

Table 1. Wasp families and genera known to harbor polydnaviruses. From Whitfield & O'Connor (2012).

A description of the life cycle and transmission process of a typical PDV can be started with the developing wasp pupa as well as at any other point in the cycle. As noted, PDVs have a transmission cycle unlike that of any other pathogen, and it has even been argued that they may not truly fit the definition of an insect virus (Whitfield & Asgari, 2003; Stoltz & Whitfield, 2009). They exist as proviral DNA integrated into the genome of the carrier wasp species (both male and female), and are "free-living" only during reproduction in female wasp ovarian calyx tissue, or after injection into the host of the carrier wasp. The injected PDV does not reproduce; it appears that some PDVs even lack the genes needed for self-assembly (Bézier et al., 2009). The PDV is at a virtual dead end, except that it allows for the reproduction of its genome in the developing wasp larva by blocking host immune responses that would otherwise encapsulate and kill it. By the same token, for most if not all carrier wasps, it is likely that successful parasitism is impossible if the polydnavirus is removed from the system: indeed this has been shown explicitly in a few systems (Edson et al., 1981; Stoltz & Guzo, 1986). The large, segmented, double-stranded PDV is excised from the wasp genome only in later pupal and adult stages, and only in female wasps (Gruber et al., 1996; Pasquier-Barre et al., 2002; Kroemer & Webb, 2006). The exact process by which this occurs has only recently been elucidated for some polydnaviruses (Annaheim & Lanzrein, 2007; Bezier et al., 2009), although, as stated previously, it is believed that some PDVs, particularly the bracoviruses, rely on the wasp host for assembly. Virus particles are assembled in the nuclei of calyx cells and accumulate in the lumen of the oviduct.

When the female wasp has emerged and is sufficiently mature, she searches for a suitable host to parasitize (Fig. 2A). During a sting event, the PDV is injected into the wasp's host along with additional proteins, toxin, and parasite eggs (Fig. 2B) (Pennachio & Strand, 2006). Once in the larval host, polydnavirus genes produce proteins that degrade the host's immune response and prevent encapsulation of the parasitoid egg or larva (Fig. 2C) (Stoltz et al. 1988). This is done in concert with other injected materials such as toxins. *Cotesia melanoscela* bracovirus (CmeBV), for example, acts in concert with wasp-produced venom on hemocytes in parasitized gypsy moth larvae (Guzo & Stoltz, 1985; Stoltz et al., 1988; Summers & Dib-Hajj 1995; Nalini et al., 2008). Among the best understood effects of PDVs on lepidopteran larvae are those occurring within the hemocytes of parasitized larvae.

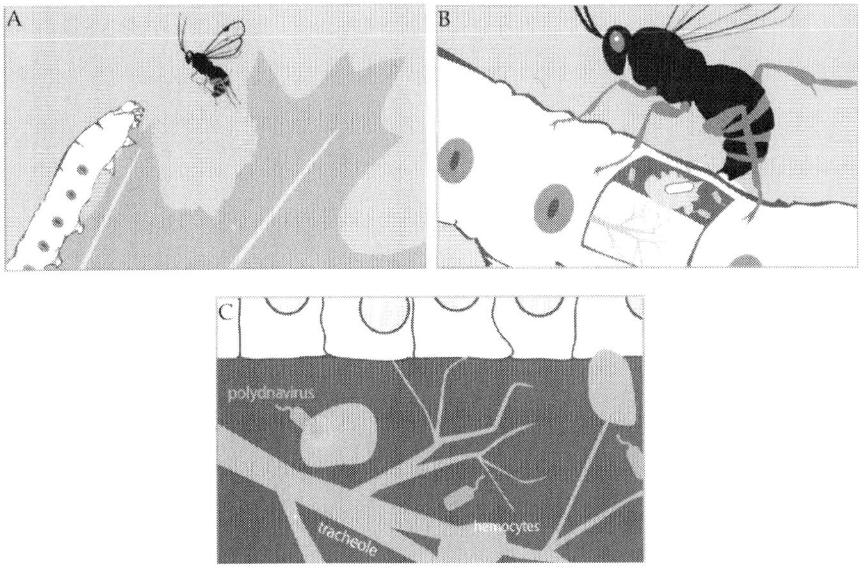

Fig. 2. Polydnavirus transmission. (A) Parasitoid wasps use series of complex cues to find and assess the condition of a host. (B) Larval Lepidoptera are injected with polydnavirus, egg, and toxin. (C) Immunosuppression of larval host occurs, primarily through PDV action on hemocytes.

If PDVs were known only for their ability to prevent encapsulation of foreign bodies, they would rightfully be the subjects of intense research. However, the immunosuppressive activity of polydnaviruses has been shown to extend to effects on baculovirus pathogenesis. This was first shown explicitly by Washburn et al. (1996), who showed that parasite injection of the *Campoletis sonorensis* ichnovirus led to immunosuppression that resulted in a more rapid spread of the AcMNPV in *H. zea* larvae. This effect was attributed to PDV-mediated abrogation of the hemocytic response as pertains to recognition of altered-self: *H. zea* is believed to clear some AcMNPV infections by encapsulating its own infected cells. This phenomenon, of special interest to the microbial ecologist, will be explored in greater detail below. So far as naturally-occurring interactions between the viruses are concerned, they are the most likely to be relevant.

4. Field interactions between polydnaviruses and nucleopolyhedroviruses

We have seen that parasitoids both produce and transmit polydnaviruses in order to suppress the immune system of their caterpillar host, and are thus able to successfully reproduce within it. In many cases baculoviruses and polydnaviruses require equivalent and possibly incompatible resources from the host, if they occur in the same host species (Caballero et al., 1991; Hochberg, 1991a). For example, LdMNPV and *C. melanoscela* are commonly encountered mortality agents of the gypsy moth in field populations in eastern United States. The timing of mortality caused by these agents overlaps completely (Woods et al., 1991; Crossman, 1922). Both enter a host during its larval stage and convert some or most of the host biomass into virus particles (baculoviruses), or aid in its conversion to a parasitoid (polydnaviruses). If a baculovirus kills or physiologically degrades a larva past a certain point, it will no longer sustain a larval parasitoid. If a parasitoid kills its host outright, as most do, baculovirus pathogenesis will be stopped at the point of host death. This would result in reduced or absent production and release of occluded virus, which is necessary to transmit the baculovirus to new hosts. Theoretically, then, a conflict arises whenever polydnaviruses and baculoviruses share a host. How may these organisms interact? Key to our exploration of this question is an analysis of the steps and barriers facing baculoviruses and polydnaviruses, and where these stages of infection overlap and are subject to interference or assistance from each other. To aid in this we will again refer to previous figures (Figs. 1, 2) mapping out both the baculovirus and polydnavirus infection processes, and dividing each process into distinct stages. At some of these stages the viruses are unlikely to influence the progression of one another, while there is great potential for them to do so at other stages. At each stage, the possibilities for interference from the other infection process will be discussed, if applicable. Where necessary for illustrative purposes we use a generalized MNPV, and a bracovirus such as those known from the braconid genus *Cotesia*. This will allow a reasonably cohesive discussion while we clearly recognize the diversity of both virus types.

4.1 Parasitoid selection of suitable host and injection of polydnavirus

Infection with a polydnavirus requires selection of a host by a wasp carrier and an attempt at parasitization of a larvae. Whether or not this selection process is affected by prior infection with a baculovirus is a question that has been explored a number of times in the past. Some studies have shown parasitoid avoidance of baculovirus infected hosts (e.g. Levin et al., 1983), while others found no effects on parasitoid preference (e.g. Sait et al., 1996). Raimo et al. (1977) showed that *C. melanoscela* readily parasitizes LdMNPV-infected larvae, at least in a laboratory setting. This particular braconid is of some importance in the literature, having also been used in some of the earliest studies of PDVs (Stoltz et al., 1976; Stoltz & Vinson, 1977; Guzo & Stoltz, 1985). In all cases, the length of time since infection began is an important component of parasitoid discrimination. For example, it is unlikely that *C. melanoscela* can detect infection before it produces overt behavioral or physiological changes, by which time the larvae in the field may be three or four days into the infection process. Indeed, in their work on the question of parasitoid discrimination using *C. melanoscela*, Versoi & Yendol (1982) felt it necessary to use infected larvae that were so moribund that they "could not right themselves when turned over".

4.2 Ingestion of polyhedra and release of virions

If full-blown infection by an NPV is likely to deter a parasitoid and thereby prevent PDV infection, a similar phenomenon can occur in the opposite direction. The first stage of infection of a lepidopteran host by an NPV is ingestion of virus-contaminated plant material. It may be argued that this process can be influenced by polydnavirus infection before it has a chance to occur. Larvae that have been stung by a carrier wasp suffer from a range of physiological effects caused by wasp-injected materials. One of these materials is the polydnavirus, and it has been shown that polydnavirus effects on hosts may occur in the presence or absence of a parasitoid egg (Strand & Dover, 1991; Beckage et al. 1994; Fathpour & Dahlman, 1995; Shelby & Webb, 1997). Most relevant of these effects is the arrested development and changes in eating habits seen in parasitized larvae; an unsurprising side effect of toxification, immunosuppression, and hormonal manipulation. The chance of baculovirus infection subsequent to parasitization will be reduced if food consumption decreases, in that the risk of infection for nucleopolyhedroviruses is directly related to the amount (leaf area) of foliage eaten (Dwyer & Elkinton, 1993; Dwyer et al., 2005), and the converse is also true. The amount of food present in the midgut could have an impact on the release of virions from the polyhedral occlusion body as well. Dissolution of the polyhedron is a pH-dependent event that occurs in the highly alkaline larval midgut. Polydnavirus infection is unlikely to *directly* influence midgut pH or the release of virions from polyhedra, but the presence or absence of plant material can do so (Keating et al., 1988; Rossiter et al., 1988). There may be no true competitive component to this aspect of polydnavirus interference with baculovirus transmission in evolutionary terms. However, it significantly influences subsequent infection risk for the parasitized larvae, and thus the parasitoid larvae as well. This is important when assessing the overall probability of both viruses occurring in the same host at the same time.

Might carrier wasps themselves increase the risk of baculovirus ingestion and infection? Parasitoid wasps have been acknowledged as having a role in the field dissemination of baculoviruses. The limited research that exists on this topic has focused particularly on the baculoviruses that infect lepidopteran hosts (Kurstak & Vago, 1967; Hochberg 1991a; 1991b; reviewed by Cossentine, 2009) for reasons of practicality and economic importance already discussed. These baculovirus infections in the field typically occur via ingestion of OBs on foliage, although Raimo et al. (1977) and others have noted that parasitoids can themslves act as a physical vectors of NPVs. These studies found that contamination of the ovipositor occurs when a wasp stings larvae, resulting in NPV infections in subsequently stung larvae. Early work (e.g. Raimo et al., 1977) did not always compare NPV-induced mortality between NPV + parasitization and NPV-only treatments, partly because such comparisons would have been of special interest only if polydnaviruses had been recognized as causing immunosuppression of the host at that time. When a nucleopolyhedrovirus is present in caterpillar larvae or on their food, parasitoid-mediated transmission of the NPVs could also occur in a number of other ways through contamination of and transfer from the parasitoid's body (Raimo et al., 1977; Young & Yearian, 1989; 1990; Hochberg, 1991a, 1991b). Yet another possibility was raised by Stoltz & Makkay (2003), who found strong circumstantial evidence that latent viruses in *T. ni* larvae could be activated after parasitism by the ichneumonid wasp *Hyposoter exiguae*, which is also known to harbor a polydnavirus (Stoltz & Makkay, 2000). None of these except for the last can be considered direct viral interactions, but they are nevertheless noteworthy.

4.3 Penetration of the peritrophic membrane

As described in Section 2, virions freed from the dissolved occlusion body must penetrate the peritrophic membrane in order to initiate infection of the insect by a baculovirus. Movement through the peritrophic membrane is thought to be facilitated by metalloproteinases termed enhancins (Slavicek, this volume). Enhancins are located within the ODV membrane in the LdMNPV (Slavicek & Popham, 2005) or are co-occluded in the granule of granuloviruses (Wang et al., 1994; Lepore et al., 1996). However, most baculoviruses lack enhancin genes; consequently other means that have yet to be elucidated are used to traverse the peritrophic membrane. Effects of ingestion-related changes to the midgut discussed in 4.2, such as changes in pH, could also influence the penetration of the membrane by virions. This hypothetical effect of gut contents or pH on the post-release penetration of the peritrophic membrane has not been explored and may be noteworthy; otherwise we feel that this stage of infection is not likely to be directly influenced by polydnavirus infection.

4.4 Spread of infection in the larval host

Once through the peritrophic membrane, NPVs infect midgut epithelial cells by binding to the cell membrane and releasing nucleocapsids into the cytoplasm. At this stage of infection, one of the strategies used by larvae to avoid infection is through apoptosis of infected midgut cells (Clarke & Clem, 2003; da Silveira et al., 2005). It has been shown that the PDV associated with *Micropletis demolitor*, the bracovirus MdBV, induces apoptosis in primary hemocytes (Strand & Pech, 1995). Recent research has shown that in some cases, however, PDV-derived proteins can inhibit baculovirus-induced apoptosis of insect cells (Kroemer & Webb, 2006), apparently via suppression of caspase activity. This is currently of interest to researchers wishing to explore methods of improving baculovirus efficacy via PDV genes. Inhibited apoptosis of midgut epithelial cells is not likely to be altering transmission probabilities in the field; however, expression of PDV has not been shown in this cell type. Budded virus leaving infected midgut cells infects other cells, and this stage is the most likely point at which the nucleopolyhedrovirus infection process could be be influenced by the action of a polydnavirus. At this point in the infection process the immune response is most likely to have a chance to clear an NPV infection, although by no means has this been shown to occur in all NPV-host systems studied.

Experiments done using injections of budded virus have produced different results. Three recent studies using budded virus injection (Rivkin et al., 2006; McNeil et al., 2010a; 2010b) all found some evidence for increased virulence when larvae were previously infected with a polydnavirus. This can be interpreted as resulting from the inability of compromised hemocytes to encapsulate foci of infection. Failed encapsulation of virus-infected cells may explain why some normally refractory larvae exhibit different pathogenesis when infected with a PDV. However, Trudeau et al. (2001) reported that a more important mechanism of resistance in *H. zea* was the unsuitability of hemocytes for infection; budded virus (BV) entered hemocytes but was unable to replicate within them, so that they served as "sinks" for BV. A similar result was obtained by Rivkin et al. (2006), who showed that the hemocytes of *S. littoralis* larvae are particularly resistant to infection by AcMNPV, and that this is an important component of resistance in this insect.

One possibility explaining these results is that unimpaired larval hemocytes could effectively encapsulate a very small number of NPV-produced foci as would be produced in the study by McNeil et al. (2010a), which used only a few budded virus particles to initiate infection, but could not encapsulate the far greater number that occur in an infection initiated *per os*. While exact numbers of foci created in typical *per os* infections are not known, analysis of budded virus production has shown that Ld652Y (*L. dispar*) cells in culture produce from approximately 15 to 125 budded virus particles per infected cell, depending on the viral isolate used (Slavicek et al., 1996; Slavicek et al., 2001). This would quickly result in an enormous number of budded virus particles in the hemocoel. The McNeil et al. (2010a) study methodology may have an analog in the field; however, if a wasp transmits NPV on a contaminated ovipositor as suggested by Raimo et al. (1977), it is likely by transferring very small amounts of budded virus from previously parasitized larvae that were already infected.

Research has been also done in granulovirus-parasitoid systems (Matthews et al., 2004; Santiago-Alvarez & Caballero, 1990), and in studies that used NPVs that are normally not found in the hosts used (Washburn et al., 2000). These are not considered here. In the experiments on normally permissive hosts (Beegle & Oatman, 1974; Raimo et al., 1977; Eller et al., 1988; Murray et al. 1995; Escribano et al., 2000; 2001) that explored polydnavirus interactions with baculovirus challenge, or can be inferred as having done so, none found clear positive effects of coinfection on baculovirus-caused mortality. All of those studies used a *per os* challenge, and are described in greater detail in Section 5 below.

5. PDV-NPV experiments using *per os* nucleopolyhedrovirus

Of primary interest to researchers of field interactions between PDVs and NPVs, are those few experiments that have used *per os* techniques to infect parasitized larvae with NPVs. This technique can be considered the most accurate representation of naturally-occurring exposure to both viruses, and we therefore have given more detailed information on these experiments below. All parasitoid names are followed by parentheses containing the name of the associated PDV.

In an early comparison of NPV mortality in parasitized and unparasitized larvae, Beegle & Oatman (1974) used the *T. ni* MNPV, the parasitoid *Hyposoter exiguae* (+HeIV), and *T. ni* caterpillar larvae. In this experiment, larvae were given a range of doses of TnMNPV *per os* immediately following parasitization. Ld_{50} values for parasitized larvae were approximately twice as high as those for unparasitized larvae (3.16×10^3 vs 1.58×10^3), and LD_{95} values five times higher (7.18×10^4 vs. 1.43×10^4). The format of this experiment can be considered a reasonable representation of field conditions: *per os* acquisition of NPV after infection with a PDV via a stinging event. Subsequent experiments will be shown to bear out such results for *per os* experiments; no increase in pathogenicity of NPVs coincident with PDVs, and typically some effect in the opposite direction. In another experiment conducted before PDVs were fully recognized, Raimo et al. (1977) explored the ability of a parasitoid wasp to physically vector an NPV. This experiment, using *C. melanoscela* (+CmBV), *L. dispar,* and LdMNPV, was not designed to look at the effects of parasitization on NPV mortality. However, in the course of the study a small amount of data was collected that allows for a comparison of virus-caused mortality of unparasitized and parasitized larvae. These larvae were reared from surface-sterilized field-collected eggs. Natural mortality caused by virus in larvae not exposed to parasites was 3.8%. Of larvae exposed to uncontaminated parasites for 2 hours and 24 hours, 9.5 % and 10.0%, respectively, died of virus. We note that this evidence is suspect, considering the possibility of contamination inherent in the use of field-collected material. We also note, however, the possibilities raised by Stoltz & Makkay (2003), of a PDV producing overt infections of a latent baculovirus. This phenomenon has been observed more often in field-collected insects than in laboratory colonies.

The study by Eller et al. (1988) compared mortality in *M. croceipes* (+McBV) parasitized and unparasitized *H. zea* larvae challenged *per os* with HzMNPV. A dose of 15,000 OBs of virus was delivered 0, 1, 2, and 4 d after parasitization (and 6, 7, 8, and 10 days after hatching). In all treatments, parasitization had no significant effects on the HzMNPV-caused mortality of larvae.

In a study that used *Chelonus insularis* (+CinsBV) to parasitize *S. frugiperda* larvae that were then challenged with SfMNPV, Escribano et al. (2000) compared SfMNPV mortality in parasitized larvae to that in virus-challenged unparasitized larvae. The LC_{50} for all instars was higher for parasitized larvae; 2nd instars (1.93×10^5 vs. 1.46×10^5 OBs/ml), 3rd instars (1.15×10^6 vs. 6.03×10^5 OBs/ml), and 4th instars (5.34×10^6 vs. 3.24×10^6 OBs/ml). The parasitized larvae in this case were exposed to *C. insularis* as eggs, which raises serious questions as to the continued presence of CinsBV in the 2nd, 3rd, and 4th instars. In Escribano et al., (2001), using the same agents as in Escribano et al. (2000), parasitized and unparasitized larvae were given the previously determined LC_{90} of SfMNPV. Although not designed to compare virus mortality, the experiment showed no significant differences in mortality between the groups (greater than 95% in both).

Three different parasitoid treatments were used in an experiment by Murray et al., (1995). In this experiment, *M. demolitor* (+MdBV), *C. kazak* (+CkBV) *H. didymator* (+HdIV), were used to parasitize larvae of *Helicoverpa armigera.* At 0 or 2 d post-parasitization these larvae were dosed with HzMNPV applied to the surface of artificial diet, in a 5-dose range from 2 to 6,000 OB/mm2 of diet. In all cases, LD_{50} values were either the same (0 d treatment) or greater (2 d treatment) for parasitized larvae.

D'Amico et al., in review. In a small experiment designed to explore the effects of *C. melanoscela* (+CmBV) parasitization on LdMNPV mortality in gypsy moth larvae, we dosed larvae *per os* with LdMNPV 2 d pre- and 2 d post-parasitization with a range of doses. In all cases, no significant differences were found between larvae that were parasitized and those that were not. We consider it unlikely that changes in CmBV treatments to longer times pre- or post- *per os* LdMNPV challenge would change the amount of LdMNPV-caused mortality in this type of experiment. Both the LdMNPV and CmeBV infection processes have passed critical stages after 48 hours or sooner, if movement of budded LdMNPV out of midgut cells (McNeil et al., 2010a) or the presence of CmeBV-coded products (Guzo & Stoltz, 1985) are considered indicators of success for these viruses in this host. Changes in timing could, however, negatively affect the success of *C. melanoscela* parasitism as the quality of the gypsy moth larval host is compromised by LdMNPV infection.

6. Conclusions

For many larval Lepidoptera, parasitism by a polydnavirus-carrying wasp or infection by a nucleopolyhedrovirus are among the top mortality factors in the field. It is also clear that by necessity, parasitoid wasps and their PDVs compete with NPVs for the larval resource. It will require focused research in this area, however, to elucidate the role played by interactions between polydnaviruses and nucleopolyhedroviruses. In the few studies that have been done, there is no obvious indication that concurrent PDV and NPV-infection leads to greater NPV mortality in lepidopteran larvae challenged with NPVs with which they are normally associated. This is true despite the widely-accepted range of immunosuppressive effects of PDV on the host insect. There is some evidence for the opposite effect, which makes good competitive sense: if a parasitized host continues to be exposed to risk of NPV infection, it is in the interest of the parasitoid to reduce the chance of host death prior to successful maturation of the parasitoid progeny.

7. Abbreviations

NPV - nucleopolyhedrovirus
OB – occlusion body
ODV – occlusion-derived virus
PO - phenoloxidase
GLD - glucose dehydrogenase
PDV - polydnavirus
BV – bracovirus
IV - ichnovirus

8. References

Annaheim, M, & Lanzrein, B. 2007. Genome organization of the *Chelonus inanitus* polydnavirus: excision sites, spacers and abundance of proviral and excised segments. J. Gen Virol., 88: 450-457.

Banks, J.C. & Whitfield, J.B. 2006. Dissecting the ancient rapid radiation of microgastrine wasp genera using additional nuclear genes. Mol. Phylogenet. Evol., 41: 690–703.

Beckage, N. B., Tan, F., Schleifer, K. W., Lane, R. D. & Cherubin, L. L., 1994. Characterization and biological effects from *Cotesia congregata* polydnavirus on host larvae of the tobacco hornworm, *Manduca sexta*. Arch. Insect. Biochem., 26: 165-195.

Beegle, C.C. & Oatman, E.R. 1974. Differential susceptibility of parasitized and nonparasitized larvae of *Trichoplusia ni* to a nuclear polyhedrosis virus. J. Invertebr. Pathol. 24: 188-195.

Belle, E., Beckage, N.E., Rousselet, J. Poirié, M., Lemeunier, F. & Drezen, J.-M. 2002. Visualization of polydnavirus sequences in a parasitoid wasp chromosome. J. Virol., 76: 5793–5796

Bézier, A, Herbinière, J, Lanzrein, B. & Drezen, J.M. 2009. Polydnavirus hidden face: The genes producing virus particles of parasitic wasps. J. Inv. Pathol., 101: 194-203

Cabellero, P., Vargas-Osuna, E. & Santiago-Alvarez, C. 1991. Parasitization of granulosis virus-infected and noninfected *Agrotis segetum* larvae and the virus transmission by three hymenopteran parasitoids. Entomol. Exp. Appl. s58: 55-60.

Clarke, T. E. & Clem, R. J. 2003. *In vivo* induction of apoptosis correlating with reduced infectivity during baculovirus infection. J. Virol., 77: 2227–2232.

Clem, R. J. 2005. The role of apoptosis in defense against baculovirus infection in insects. Curr. Top. Microbiol. Immunol., 289: 113–129.

Cook, D.I. & Stoltz, D.B., 1983. Comparative serology of viruses isolated from ichneumonid parasitoids. Virology 130: 215–220.

Cory, J.S. & Myers, J.H., 2003. The ecology and evolution of insect baculoviruses, Annu. Rev. Ecol. Syst. 34: 239–272.

Cossentine, J.E. 2009. The parasitoid factor in the virulence and spread of lepidopteran baculoviruses. Virologica Sinica 24: 305-314.

Cox-Foster, D.L. & Stehr, J.E. 1994. Induction and localization of FAD-glucose dehydrogenase (GLD) during encapsulation of abiotic implants in *Manduca sexta* larvae. Journal of Insect Physiology 40: 235–249.

Crossman, S. S., 1922. *Apanteles melanoscelus*, an imported parasite of the gipsy [sic] moth. USDA Bulletin 1028, Washington, D. C.

Crouch, E.A., Cox, L.T., Morales, K.G. & Passarelli, A.L. 2007. Inter-subunit interactions of the *Autographa californica* M nucleopolyhedrovirus RNA polymerase. Virology 367: 265–274.

da Silveira, E. B., Cordeiro, B. A., Ribeiro, B. M. & Bao, S. N. 2005. In vivo apoptosis induction and reduction of infectivity by an *Autographa californica* multiple nucleopolyhedrovirus p352 recombinant in hemocytes from the velvet bean caterpillar *Anticarsia gemmatalis* (Lepidoptera: Noctuidae). Res Microbiol 156: 1014–1025.

d'Alencon, E., Piffanelli, P., Volkoff, A.N., Sabau, X., Gimenez, S., Rocher, J., Cerutti, P. & Fournier, P. 2004. A genomic BAC library and a new BAC-GFP vector to study the holocentric pest *Spodoptera frugiperda*. Insect Biochem Mol Biol. 34: 331–41.

David, W.A.L. 1975. The status of viruses pathogenic for insects and mites. Annu. Rev. Entomol. 20: 97-117.

Dougherty E.M., Narang N., Loeb M., Lynn D.E. & Shapiro M. 2006. Fluorescent brightener inhibits apoptosis in baculovirus-infected gypsy moth larval midgut cells. Biocontrol Sci. Technol. 16: 157–168

Dwyer, G. & J.S. Elkinton. 1993. Using simple models to predict virus epizootics in gypsy-moth populations. J. of Anim. Ecol., 62: 1-11.

Dwyer, G., Firestone J., & Stevens T.E. 2005. Should models of disease dynamics in herbivorous insects include the effects of variability in host-plant foliage quality? Am. Nat., 165: 16-31.

Edson, K.M., Vinson, S.B., Stoltz, D.B. & Summers M.D. 1981. Virus in a parasitoid wasp: suppression of the cellular immune response in the parasitoid's host. Science 211: 582-583.

Eller, F. J., Boucias, D. G. & Tumlinson, J. H. 1988. Interactions between *Microplitis croceipes* (Hymenoptera:Braconidae) and a nuclear polyhedrosis virus of *Heliothis zea* (Lepidoptera: Noctuidae). Environ. Entomol., 17: 977- 982.

Engelhard, E. K. & Volkman, L. E. 1995. Developmental resistance in 4th instar *Trichoplusia ni* orally inoculated with *Autographa californica* M nuclear polyhedrosis virus. Virology 209: 384–389.

Engelhard, E. K., Kammorgan, L. N. W., Washburn, J. O. & Volkman, L. E. 1994. The insect tracheal system: a conduit for the systemic spread of *Autographa californica* M nuclear polyhedrosis virus. Proc Natl Acad Sci USA 91: 3224–3227.

Erlandson, M. 2008. Insect pest control by viruses. Encyclopedia of Virology, Third Edition, 3: 125-133.

Escribano, A., Williams, T., Goulson, D., Cave, R.D. & Caballero, P., 2000. Parasitoid–pathogen–pest interactions of *Chelonus insularis, Campoletis sonorensis,* and a nucleopolyhedrovirus in *Spodoptera frugiperda* larvae. Biol. Control, 19: 265–273.

Escribano, A., Williams, T., Goulson, D., Cave, R.D., Cha peritrophic membrane an, J.W. & Caballero, P. 2001. Consequences of interspecific competition on the virulence and genetic composition of a nucleopolyhedrovirus in *Spodoptera frugiperda* larvae parasitized by *Chelonus insularis*. Biocontrol, 11: 649-662.

Fang, M., Nie, Y., Harris, S., Erlandson, M.A. & Theilmann, D.A. 2009. *Autographa californica* multiple nucleopolyhedrovirus core gene ac96 encodes a *per os* infectivity factor (PIF-4). J. Virol. 83: 12569–12578.

Fang, M., Nie, Y., Wang, Q., Deng, F., Wang, R., Wang, H., Wang, H., Vlak, J.M., Chen, X. & Zu, Z. 2006. Open reading frame 132 of *Helicoverpa armigera* nucleopolyhedrovirus encodes a functional per os infectivity factor (PIF-2). J. Gen Virol. 87: 2563–2569.

Fathpour, H. & Dahlman, D.L. 1995. Polydnavirus of *Microplitis croceipes* prolongs the larval period and changes hemolymph protein content of the host, *Heliothis virescens*. Arch Insect Biochem Physiol 28: 33–48.

Faulkner, P., Kuzio, J., Williams, G.V. & Wilson, J.A. 1997. Analysis of p74, a PDV envelope protein of Autographa californica nucleopolyhedrovirus required for occlusion body infectivity in vivo. J. Gen. Virol. 78: 3091–3100.

Fleming, J.G.W. & Summers, M.D., 1991. Polydnavirus DNA is integrated into the DNA of its parasitoid host wasp. Proc. Natl. Acad. Sci. 88: 9770-9774.

Fleming, J.G.W. & Summers, M.D., 1986. *Campoletis sonorensis* endoparasitic wasps contain forms of *C. sonorensis virus* DNA suggestive of integrated and extrachromosomal polydnavirus DNAs. J. Virol. 57: 552–562.

Flipsen, J.T., Martens, J.W., vanOers, M.M., Vlak, J.M. & van Lent, J.W. 1995. Passage of *Autographa californica* nuclear polyhedrosis virus through the midgut epithelium of *Spodoptera exigua* larvae. Virology 208: 328–35.

Granados, R.R. & Lawler, K.A. 1981. In vivo pathway of *Autographa californica* baculovirus invasion and infection. Virology 108: 297-308.

Gruber, A., Stettler, P., Heineger, P., Schumpli, D., & Lanzrein, B. 1996. Polydnavirus DNA of the braconid wasp *Chelonus inanitus* is integrated in the wasp genome and excised only in later pupal and adult stages of the female. J. Gen. Virol. 77, 2873-2879.

Guzo, D. & Stoltz, D. B., 1985. Obligatory multiparasitism in the tussock moth, *Orgyia leucostigma*. Parasitology 90, 1-10.

Haas-Stapleton, E. J., Washburn, J. O. & Volkman, L. E. 2003. Pathogenesis of *Autographa californica* M nucleopolyhedrovirus in fifth instar *Spodoptera frugiperda*. J. Gen Virol 84: 2033–2040.

Harrap, K.A. & Payne, C.C. 1979. The structural properties and identification of insect viruses. Adv. Virus Res. 25:273-355.

Harrison R.L., Sparks W.O. & Bonning B.C. 2010. *Autographa californica* multiple nucleopolyhedrovirus ODV-E56 envelope protein is required for oral infectivity and can be substituted functionally by *Rachiplusia ou* multiple nucleopolyhedrovirus ODV-E56. J. Gen Virol. 91:1173–1182.

Hegedus, D., Erlandson, M., Gillott, C. & Toprak, U. 2009. New insights into peritrophic matrix synthesis, architecture, and function. Annu Rev Entomol. 54: 285–302.

Herniou, E.A. & Jehle, J.A. 2007. Baculovirus phylogeny and evolution. Curr. Drug Targets 8: 1043-50.

Hochberg, M.E., 1991a. Intra-host interactions between a braconid endoparasitoid, *Apanteles glomeratus*, and a baculovirus for larvae of *Pieris brassicae*. J. Anim. Ecol., 60: 51-63.

Hochberg, ME. 1991b. Extra-Host interactions between a braconid endoparasitoid, *Apanteles glomeratus*, and a baculovirus for larvae of *Pieris brassicae* J. Anim. Ecol. 60: 65-77.

Hoover, K., Grove, M. J. & Su, S. Z. 2002. Systemic component to intrastadial developmental resistance in *Lymantria dispar* to its baculovirus. Biol Control 25: 92–98.

Horton H.M. & Burand J.P. 1993. Saturable attachment sites for polyhedron-derived baculovirus on insect cells and evidence for entry via direct membrane fusion. J. Virol. 67: 1860–1868.

Jehle, J.A., Lange, M., Wang, H., Zhihong, H., Wang, Y. & Hauschild, R. 2006. Molecular identification and phylogenetic analysis of baculoviruses from Lepidoptera. Virology 346: 180-193.

Jiravanichpaisal, P., Lee, B.L. & Soderhall, K. 2006. Cell-mediated immunity in arthropods: hematopoiesis, coagulation, melanization and opsonization. Immunobiology 211: 213–236.

Keating, S. T., Yendol, W.G. & Schultz, J.C.. 1988. Relationship between susceptibility of gypsy moth larvae (Lepidoptera: Lymantriidae) to a baculovirus and host plant foliage constituents. Environmental Entomology 17: 952-958.

Keddie B.A., Aponte G.W. & Volkman L.E. 1989. The pathway of infection of *Autographa californica* nuclear polyhedrosis virus in an insect host. Science. 243: 1728–1730.

Kikhno, I., Gutierrez, S., Croizier, L., Croizier, G. & Ferber, M.L. 2002. Characterization of pif, a gene required for the per os infectivity of *Spodoptera littoralis* nucleopolyhedrovirus. J. Gen Virol. 83: 3013–3022.

Kroemer, J.A. & Webb, B.A. 2006 Divergences in protein activity and cellular localization within the *Campoletis sonorensis* ichnovirus Vankyrin family. J. Virol. 80: 12219–12228.

Kurstak, E. & Vago, C., 1967. Transmission dun virus de la densonucleose par la parasitisma d'un hymenoptere. Rev. Can. Biol. 26: 311-316.

LaPointe, R., Tanaka, K., Barney, W.E., Whitfield, .JB., Banks, J.C., Beliveau, C., Stoltz, D., Webb, B.A. & Cusson, M. 2007. Genomic and morphological features of a banchine polydnavirus: Comparison with Bracoviruses and Ichnoviruses. J. Virol 81: 6491–9501.

Lee,M., Yoon, C.S., Yi, J., Cho, J.R. & Kim, H.S. 2005. Cellular immune responses and FAD-glucose dehydrogenase activity of *Mamestra brassicae* Lepidoptera: Noctuidae) challenged with three species of entomopathogenic fungi. Physiological Entomology 30: 287–292.

Lepore, L. S., Roelvink, P.R. & Granados, R.R. 1996. Enhancin, the granulosis virus protein that facilitates nucleopolyhedrovirus (NPV) infections, is a metalloprotease. *J. Invertebr. Pathol.* 68: 131-140.

Levin, D. B., Laing, J. E., Jaques, R. P. & Corrigan, J. E., 1983. Transmission of the granulosis virus of *Pieris rapae* (Lepidoptera: Pieridae) by the parasitoid *Apanteles glomeratus* (Hymenoptera: Braconidae). Environ. Entomol. 12: 166–170.

Li, X., Song, J., Jiang, T., Liang, C. & Chen, X. 2007. The N-terminal hydrophobic sequence of Autographa californica nucleopolyhedrovirus PIF-3 is essential for oral infection. Arch Virol. 152: 1851–1858.

Lovallo, N. & Cox-Foster, D.L. 1999. Alteration in FAD-glucose dehydrogenase activity and hemocyte behavior contribute to initial disruption of *Manduca sexta* immune response to *Cotesia congregata* parasitoids. Journal of Insect Physiology 45: 1037–1048.

Lovallo, N.C., McPherson, B.A. & Cox-Foster, D.L., 2002. Effects of the polydnavirus of *Cotesia congregata* on the immune system and development of non-habitual hosts of the parasitoid, J. Insect Physiol. 48: 517–526.

Matthews H J, Smith I, & Bell H A. 2004. Interactions between the parasitoid *Meteorus gyrator* (Hymenoptera:Braconidae) and a granulovirus in *Lacanobia oleracea* (Lepidoptera: Noctuidae). Environ. Entomol., 33: 949-957.

McNeil, J., Cox-Foster, D., Slavicek, J. & Hoover, K., 2010a. Contributions of immune responses to developmental resistance in *Lymantria dispar* challenged with baculovirus. J. Insect Physiol., 56: 1167-1177.

McNeil, J., Cox-Foster, D., Gardner, M., Slavicek, J., Thiem, S, & Hoover, K. 2010b. Pathogenesis of *Lymantria dispar* multiple nucleopolyhedrovirus in *L. dispar* and mechanisms of developmental resistance. J. Gen. Virol. 91: 1590-1600.

Murphy, N., Banks, J.C., Whitfield, J.B. & Austin, A.D., 2008. Phylogeny of the microgastroid complex of subfamilies of braconid parasitoid wasps (Hymenoptera) based on sequence data from seven genes, with an improved estimate of the time of origin of the lineage. Mol. Phylogen. Evol. 47: 378–395.

Moscardi, F. 1999. Assessment of the application of baculoviruses for control of Lepidoptera. Annu. Rev. Entomol. 44: 257-289.

Murray, D.A., Monsour, C.J. & Teakle R.E., 1995. Interactions between nuclear polyhedrosis virus and three larval parasitoids of *Helicoverpa armigera* (Hübner). J. Aust. Ent. Soc., 34: 319-322.

Nalini, M. et al., Choi, J.Y., Je, Y.H., Hwang, I. & Kim, Y., 2008. Immunoevasive property of a polydnaviral product, CpBV-lectin, protects the parasitoid egg from hemocytic encapsulation of *Plutella xylostella* (Lepidoptera: Yponomeutidae). J. Insect Physiol. 54: 1125-1131.

Nappi, A.J. & Christensen, B.M. 2005. Melanogenesis and associated cytotoxic reactions: Applications to insect innate immunity. Insect Biochemistry and Molecular Biology 35: 443–459.

Ohkawa, T., Volkman, L.E. & Welch M.D. 2010. Actin-based motility drives baculovirus transit to the nucleus and cell surface. J. Cell Biol.190: 187–95.

Ohkawa, T., Washburn,J.O., Sitapara, R., Sid, E. & Volkman, L.E. 2005. Specific binding of *Autographa californica* M nucleopolyhedrovirus occlusion-derived virus to midgut cells of *Heliothis virescens* larvae is mediated by products of pif genes Ac119 and Ac022 but not by Ac115. J. Virol. 79: 15258–15264.

Pasquier-Barre, F., Dupuy, C., Huguet, E., Moneiro, F. & Moreau, A. 2002. Polydnavirus replication: The EP1 segment of the parasitoid wasp *Cotesia congregata* is amplified within a larger precursor molecule. J. Gen. Virol. 83: 2035–45

Peng, K., van Oers, M.M., Hu, Z., van Lent, J.W. & Vlak, J.M. 2010. Baculovirus per os infectivity factors form a complex on the surface of occlusion-derived virus. J. Virol. 84: 9497–9504.

Pennacchio, F. & Strand, M. R. 2006. Evolution of developmental strategies in parasitic Hymenoptera. Annu. Rev. Entomol. 51: 233 -258.

Pritchett, D. W., Young, S. Y. & Yearian, W. C. 1984. Some factors involved in the dissolution of *Autographa californica* nuclear polyhedrosis virus polyhedra by digestive fluids of *Trichoplusia ni* larvae. J. Invertebr. Pathol. 43: 160–168.

Raimo, B., Reardon R. C. & Podgwaite, J. D., 1977. Vectoring gypsy moth nuclear polyhedrosis by *Apanteles melanoscelus*. Entomophaga, 22: 207-215.

Reardon, R. C. 1996. Gypchek, the Gypsy Moth Nucleopolyhedrosis Virus Product. Morgantown, WV: USDA Forest Service, Northeastern Area, Forest Health Technology Enterprise Team.

Riegel, C. I. & Slavicek, J. M. 1997. Characterization of the replication cycle of the Lymantria *dispar* nuclear polyhedrosis virus. Virus Res. 51: 9–17.

Rivkin, H., Kroemer, J.A., Bronshtein, A., Belausov, E., Webb, B.A. & Chejanovsky, N. 2006. Response of immunocompetent and immunosuppressed *Spodoptera littoralis* larvae to baculovirus infection. J. Gen. Virol. 87: 2217–2225.

Rohrmann, G.F., 2011. Baculovirus Molecular Biology. National Library of Medicine (US), NCBI, Available at [http://www.ncbi.nlm.nih.gov/bookshelf/br.fcgi?book=bacvir].

Rossiter, M. C., Schultz, J.C. & Baldwin, I.T. 1988. Relationships among defoliation, red oak phenolics and gypsy moth growth and reproduction. Ecology 69: 267-277.

Rotheram, S.M. 1973. The surface of the egg of a parasitic insect. II. The ultrastructure of the particulate coat on the egg of *Nemeritis*. Proc R Soc Lond Series B, 183: 195 - 204.

Sait, S. M., Begon, M., Thompson, D. J. & Harvey, J. A., 1996. Parasitism of baculovirus-infected *Plodia interpunctella* by *Venturia canescens* and subsequent virus transmission. Funct. Ecol. 10: 586-591.

Salt, G. 1965. Experimental Studies in Insect Parasitism. XIII. The Haemocytic Reaction of a Caterpillar to Eggs of its Habitual Parasite. Proc. Royal Society of London. Series B, 162: 303-318.

Salt, G. 1970. Cellular defense reactions of insects. Cambridge University Press, London and New York.

Santiago-Alvarez, C. & Caballero, P., 1990. Susceptibility of parasitized *Agrotis segetum* larvae to a granulosis virus. J. Invertebr. Pathol. 56: 128-131.

Schmidt, O., Theopold, U. & Strand, M.R., 2001. Innate immunity and its evasion and suppression by hymenopteran endoparasitoids. BioEssays 23: 344–351.

Shelby, K.S., & Webb, B.A. 1997. Polydnavirus infection inhibits translation of specific growth-associated host proteins. Insect Biochem. Mol. Biol., 27: 263–270.

Shelby, K.S. & Popham, H.J.R. 2006. Plasma phenoloxidase of the larval tobacco budworm, *Heliothis virescens*, is virucidal. J. of Insect Sci. 6: 13.

Shrestha, S. & Kim, Y.G. 2008. Eicosanoids mediate prophenoloxidase release from oenocytoids in the beet armyworm *Spodoptera exigua*. Insect Biochem. Mol. Biol. 38: 99–112.

Slavicek, J. M., Mercer, M. J., Kelly, M. E. & Hayes-Plazolles, N., 1996. Isolation of a baculovirus variant that exhibits enhanced polyhedra production stability during serial passage in cell culture. J. Invertebr. Pathol. 67: 153–160.

Slavicek & Popham. 2005. The *Lymantria dispar* nucleopolyhedrovirus enhancins are components of occlusion-derived virus. J. Virol. 79: 10578-10588.

Slavicek, J.M., Hayes-Plazolles, N. & Kelly, M.E., 2001. Identification of a *Lymantria dispar* nucleopolyhedrovirus isolate that does not accumulate few-polyhedra mutants during extended serial passage in cell culture. Biol. Con., 22: 159-168.

Stanley, D. & Shapiro, M. 2007. Eicosanoid biosynthesis inhibitors increase the susceptibility of *Lymantria dispar* to nucleopolyhedrovirus LdMNPV. Journal of Invertebrate Pathology 95: 119–124.

Stanley, D.W. & Shapiro, M. 2009. Eicosanoids influence insect susceptibility to nucleopolyhedroviruses. Journal of Invertebrate Pathology 102: 245–249.

Stoltz, D.B., Vinson, S.B. & Mackinnon, E.A., 1976. Baculovirus-like particles in the reproductive tracts of female parasitoid wasps. Can. J. Microbiol. 27: 1013–1023.

Stoltz, D. B. & Vinson, S. B., 1977. Baculovirus-like particles in the reproductive tracts of female parasitoid wasps. II. The genus *Apanteles*. Can. J. Microbiol. 23: 28-37.

Stoltz, D.B. & Vinson, S.B., 1979. Viruses and parasitism in insects. Advances in Virus Research 24: 125–171.

Stoltz, D. B., Krell, P.J. & Vinson, S.B. 1981. Polydisperse viral DNAs in ichneumonid ovaries: a survey. Can. J. Microbiol. 27: 123-130.

Stoltz, D. B., Guzo, D. & Cook, D. 1986. Studies on polydnavirus transmission. Virology 155: 120-131.

Stoltz, D. B. & Guzo, D., 1986. Apparent haemocytic transformations associated with parasitoid-induced inhibition of immunity in *Malacosoma disstria* larvae. J. Insect Physiol. 32: 377-388.

Stoltz, D.B., Guzo D., Belland, E.R., Lucarotti, C.J. & MacKinnon, E.A., 1988. Venom promotes uncoating in vitro and persistence in vivo of DNA from a braconid polydnavirus. J. Gen. Virol. 69: 903-907.

Stoltz, D. B. 1990. Evidence for chromosomal transmission of polydnavirus DNA. J. Gen. Virol. 71: 1051-1056.

Stoltz, D. B. & Whitfield, J. B., 1992. Viruses and virus-like entities in the parasitic Hymenoptera. J. Hymenopt. Res. 1: 125–139.

Stoltz, D. & Makkay, A., 2000. Co-replication of a reovirus and a polydnavirus in the ichneumonid parasitoid, *Hyposoter exiguae*. Virology 278: 266-275,.

Stoltz, D. & Makkay, A., 2003. Overt viral diseases induced from apparent latency following parasitization by the ichneumonid wasp, *Hyposoter exiguae*. J. Insect Phys. 49: 483-489.

Stoltz, D.B. & Whitfield, J.B., 2009. Virology. Making nice with viruses. Science 323: 884–885.

Strand, M. R. & Dover, B. A. 1991. Developmental disruption of *Pseudoplusia includens* and *Heliothis virescens* larvae by the calyx fluid and venom of *Microplitis demolitor*. Arch. Insect Biochem. Physiol. 18: 131-145.

Strand, M.R. & Pech, L.L. 1995. *Microplitis demolitor* polydnavirus induces apoptosis of a specific haemocyte morphotype in *Pseudoplusia includens*. J. Gen. Virol. 76: 283-291

Summers, M.D. & Dib-Hajj., S.D. 1995. Polydnavirus-facilitated endoparasite protection against host immune defenses. Proc. Natl. Acad. Sci., 92: 29-36.

Szewczyk, B., Hoyos-Carvajal, L., Paluszek, M., Skrzecz, I. & Souza, M.L. 2006. Baculovirus - re-emerging biopesticides. Biotechnol. Adv. 24: 143-160.

Szewczyk, B., Rabalski, L., Krol, E., Sihler, W., & Souza, M.L. 2009. Baculovirus biopesticides - a safe alternative to chemical protection of plants. J. Biopestic. 2: 209-216.

Tinsley, T.W. & Harrap, K.A. 1978. Viruses of invertebrates. Compr. Virol. 12: 1-101.

Trudeau, D., Washburn, J. O. & Volkman, L. E. 2001. Central role of hemocytes in *Autographa californica* M nucleopolyhedrovirus pathogenesis in *Heliothis virescens* and *Helicoverpa zea*. J. Virol. 75: 996–1003.

Turnbull, M.W. & Webb, B.A., 2002. Perspectives on polydnavirus origin and evolution. Adv. Virus Res. 58: 203- 254.

Versoi, P.L. & Yendol, W.G., 1982. Discrimination by the parasite, *Apanteles melanoscelus*, between healthy and virus-infected gypsy moth larvae. Environ. Entomol. 11, 42-45.

Volkman, L. E. 2007. Baculovirus infectivity and the actin cytoskeleton. Curr Drug Targets 8, 1075–1083.

Wang, P., Hammer, D.A., & Granados, R.R. 1994. Interaction of *Trichoplusia ni* granulosis virus-encoded enhancin with the midgut epithelium and peritrophic membrane of four lepidopteran insects. J. Gen. Virol., 75: 1961–1967.

Washburn J.O., Haas-Stapleton E.J., Tan F.F., Beckage N.E. & Volkman L.E., 2000. Co-infection of *Manduca sexta* larvae with polydnavirus from *Cotesia congregata* increases susceptibility to fatal infection by *Autographa californica* M Nucleopolyhedrovirus. J. of Insect Physiol. 46: 179–190.

Washburn, J. O., Kirkpatrick, B. A. & Volkman, L.E., 1996. Insect protection against viruses. Nature 383: 767.

Washburn, J. O., Haas-Stapleton, E. J., Tan, F. F., Beckage, N. E. & Volkman, L. E. 2000. Co-infection of *Manduca sexta* larvae with polydnavirus from Cotesia congregata increases susceptibility to fatal infection by *Autographa californica* M nucleopolyhedrovirus. J. Insect Physiol. 46: 179–190.

Webb, B. A. & Strand, M. R. 2005. The biology and genomics of polydnaviruses, in Gilbert, I., Iatrou, K., Gill S. (Eds.) Comprehensive Molecular Insect Science, San Diego, CA: Elsevier pp. 260–323.

Whitfield, J. B. 1997. Molecular and morphological data suggest a single origin of the polydnaviruses among braconid wasps. Naturwissenschaften 84: 502–507.

Whitfield, J.B., & Asgari, S. 2003. Virus or not? Phylogenetics of polydnaviruses and their wasp carriers. J. Insect Physiol., 49: 397–405.

Whitfield, J.B. & O'Connor, J.M. 2012. Molecular systematics of wasp and polydnavirus genomes and their coevolution. *In* Parasitoid Viruses: Symbionts and Pathogens, edited by Nancy E. Beckage, Jean-Michel Drezen, Elsevier Inc., in press.

Woods, S., Elkinton, J.S., Murray, K.D., Liebhold, A.M., Gould, J.R., & Podgwaite, J.D., 1991. Transmission dynamics of a nuclear polyhedrosis virus and predicting mortality in gypsy moth (Lepidoptera: Lymantriidae) populations. J. Econ. Entomol., 84: 423-430.

Wyler, S. & Lanzrein, B., 2003. Ovary development and polydnavirus morphogenesis in the parasitic wasp *Chelonus inanitus*. II. Ultrastructural analysis of calyx cell development, virion formation and release. J. Gen. Virol. 84: 1151–1163.

Young, S.Y. & Yearian W.C., 1989. Nuclear polyhedrosis virus transmission by *Microplitis croceipes* (Hymenoptera:Braconidae) adult females reared in infected *Heliothis virescens* (Lepidoptera: Noctuidae) larvae. J. Entomol. Sci., 24: 500-506.

Young, S.Y. & Yearian W.C., 1990. Transmission of nuclear polyhedrosis virus by the parasitoid *Microplitis croceipes* (Hymenoptera: Braconidae) to *Heliothis virescens* (Lepidoptera: Noctuidae) on soybean. Environ. Entomol., 19: 251-256.

Zhang, P., Yang, K., Dai, X. J., Pang, Y. & Su, D. M. 2002. Infection of wild-type *Autographa californica* multicapsid nucleopolyhedrovirus induces in vivo apoptosis of *Spodoptera litura* larvae. J. Gen. Virol. 83: 3003–3011.

9

Baculovirus Enhancins and Their Role in Viral Pathogenicity

James M. Slavicek
USDA Forest Service
USA

1. Introduction

Baculoviruses are a large group of viruses pathogenic to arthropods, primarily insects from the order Lepidoptera and also insects in the orders Hymenoptera and Diptera (Moscardi 1999; Herniou & Jehle, 2007). Baculoviruses have been used to control insect pests on agricultural crops and forests around the world (Moscardi, 1999; Szewczk et al., 2006, 2009; Erlandson 2008). Efforts have been ongoing for the last two decades to develop strains of baculoviruses with greater potency or other attributes to decrease the cost of their use through a lower cost of production or application. Early efforts focused on the insertion of foreign genes into the genomes of baculoviruses that would increase viral killing speed for use to control agricultural insect pests (Black et al., 1997; Bonning & Hammock, 1996). More recently, research efforts have focused on viral genes that are involved in the initial and early processes of infection and host factors that impede successful infection (Rohrmann, 2011). The enhancins are proteins produced by some baculoviruses that are involved in one of the earliest events of host infection. This article provides a review of baculovirus enhancins and their role in the earliest phases of viral infection.

2. Lepidopteran specific baculoviruses

The *Baculoviridae* are divided into four genera: the Alphabaculovirus (lepidopteran-specific nucleopolyhedroviruses, NPV), Betabaculovirus (lepidopteran specific Granuloviruses, GV), Gammabaculovirus (hymenopteran-specific NPV), and Deltabaculovirus (dipteran-specific NPV) (Jehle et al., 2006). Baculoviruses are arthropod-specific viruses with rod-shaped nucleocapsids ranging in size from 30-60 nm x 250-300 nm. Most baculoviruses produce two types of virus particles, the occlusion-derived virion (ODV) and the budded virion (BV). ODV are enclosed in a paracrystalline protein matrix, termed the occlusion body (OB) produced by NPVs and the granule produced by GVs, which are composed primarily of the proteins polyhedrin and granulin, respectively. The OB/granule provides the embedded ODV protection from environmental elements such UV light, rain, etc. The BV is produced in the early stages of viral replication and spreads the infection throughout the susceptible cells within the host (Rohrmann, 2011). The NPVs produce OBs that range in size from 0.15 to 3 μm in size and contain many virions. The GVs generate smaller ovoid shaped granules of 0.13 - 0.5 μm that contain a single virion (Ackermann & Smirnoff, 1983). Baculovirus genomes are circular double-stranded DNA molecules ranging in length from about 80 -180 kbp and contain from approximately 90 to 185 open reading frames (ORFs). All baculoviruses sequenced to date contain unique and common ORFs. More than 800 orthologous gene groups have been identified as a consequence of speciation events during viral evolution (Jehle, et. al., 2006). Baculovirus genomes undergo a high rate of mutation as a consequence of gene duplication and loss, homologous recombination, and gene transfer from other viruses, bacteria, and eukaryotic genomes. Sequence analysis of NPV and GV genomes revealed that the NPVs and GVs contain many gene homologs, and 29 core genes are conserved in all baculoviruses sequenced to date (Herniou et al., 2003; Jehle et al., 2006; Herniou & Jehle, 2007). Several groups of genes are conserved in some but not all NPVs and GVs. One of these groups is termed auxiliary genes, which include *enhancin* genes. These genes are not essential for viral DNA replication but provide a selective advantage to a virus (Miller, 1997).

3. Viral pathogenesis

GVs and NPVs initiate infection when a susceptible host ingests viral OBs present on host plants. Within the alkaline environment of the larval midgut, OBs dissolve, thereby releasing ODV. ODV must first traverse a physical structure termed a peritrophic matrix (PM), which is composed of proteins, mucopolysaccharides, and chitin (Pritchett et al., 1984; Hegedus, et al., 2009). The PM provides a barrier to gut cells from bacteria, viruses, fungi, and physical damage from ingested plant material (Lehane, 1997; Terra, 2001). The PM is in a constant state of regeneration from epithelial cells as the larvae feeds. The movement of food material through the insect gut causes loss of the PM.

ODV then gains entry into midgut cells by a type of fusion process (Granados, 1980; Granados & Lawler, 1981) that is not defined. The type NPV, *Autographa californica* multiple NPV (AcMNPV), initiates the infection cycle by infecting primarily columnar epithelial cells within the midgut and to a much lesser extent regenerative epithelial cells in *Trichoplusia ni* (Keddie et al., 1989) or *Spodoptera exigua* larvae (Flipsen et al., 1995). Several factors are involved during the initial act of infection that includes ODV binding to midgut cells at cell receptors and viral entry into the cells. All sequenced baculoviruses contain genes that code for *per os* infectivity factors (PIFs) that are associated with ODVs but not BV (Faulkner et al., 1997; Kikhno et al., 2002; Fang et al., 2006; Harrison et al., 2010; Fang et al., 2009). The *pif* genes include p74-pif, and *pif* genes 1-5, Ac119, Ac 22, Ac115, Ac96, and Ac148, respectively. Deletion of any of the *pif* genes from a viral genome significantly decreases but does not eliminate *per os* infectivity (d'Alencon et al., 2004; Crouch et al., 2007). The PIFs, with the exception of PIF3, are thought to be involved in binding or interacting with the midgut cells that leads to infection (Ohkawa, et al., 2005; Li et al., 2007; Peng et al., 2010; Horton & Burand, 1993). Another gene present in some GVs and NPVs, the *enhancin* gene, codes for a protein that also impacts viral potency during *per os* infection as described in the next section.

Upon entry into midgut cells, the nucleocapsids are actively transported to nuclear pores in a process that uses actin polymerization (Ohkawa et al., 2010). Viral DNA is then released into the cell nucleus and viral replication ensues (Rohrmann, 2011). During the early phase of viral replication, BV are produced that bud from midgut cells and infect tracheal epidermal cells, which penetrate the basal lamina (Volkman, 2007). Infection spreads via the tracheal system and haemocytes until many different cell types are infected (Engelhard et al., 1994). During the later phase of viral replication, ODV are produced and packaged within OBs. Upon the host's death, liquefaction occurs, releasing OBs into the environment that can lead to infection of another host.

4. Early studies on the "synergistic factor" in Granuloviruses

Early studies with GVs identified a factor in the *Pseudaletia unipuncta* (Psun) GV that increased the infectivity of PsunNPV in NPV/GV mixed infections (Tanada, 1959). Early *P. unipuncta* instars were highly susceptible to PsunNPV and PsunGV, whereas the later instars (4-6) were increasingly less susceptible. In contrast, when larvae were infected *per os* with both NPV and GV viruses, they were highly susceptible. Feeding of larvae first with PsunNPV, followed by PsunGV did not generate synergy, whereas synergy was observed when the larvae were first infected with PsunGV followed by PsunNPV indicating the synergistic factor was associated with PsunGV. The results of heat-inactivation experiments indicated that a component within the GV ODV envelope or the granule was responsible for the synergism (Tanada, 1959). The enhancing factor was named the synergistic factor (SF), and was found to be a protein component of the GV capsule (Hara et al., 1976; Tanada et al., 1973), comprising about 5% of the capsule protein components (Yamamoto & Tanada, 1978). When purified SF was added to PsunNPV ODV it exhibited a strong affinity for viral envelopes, and with about 8 molecules of SF bound to each enveloped virion the infectivity of the ODV were significantly enhanced (Yamamoto & Tanada, 1980). A synonymous factor, (viral enhancing factor, VEF) was found in *Trichoplusia ni* GV (TnGV) and *Xestia c-nigrum* GV (XecnGV) granules, which enhanced the infectivity of AcNPV, and *Xestia c-nigrum* NPV (XecnNPV), respectively (Derksen & Granados, 1988; Gallo et al., 1991; Goto et al., 1990).

Initial studies on the function of enhancins suggested that the site of SF action is the cellular membrane of the midgut cell microvilli (Tanada et al., 1975; Tanada et al., 1980), and these cells contain specific binding sites for enhancins (Uchima et al., 1988). Electron microscopy was used to visualize and count attached virions on midgut epithelium in *P. unipuncta* larvae. Electron micrographs of midguts treated with PsunNPV and SF exhibited 40 times more nucleocapsids attached or within microvilli compared to midguts treated only with PsunNPV (Tanada et al., 1975). In addition, antibody studies also localized the site of SF binding to midgut cell membranes (Tanada et al., 1980). Competition binding studies with Concanavalin A and castor bean lectin were found to inhibit binding of SF to midgut cell membranes, suggesting that SF binding was to specific receptors (Uchima et al., 1988).

Subsequent studies by Granados and colleagues demonstrated that the enhancin from TnGV degraded major glycoproteins of 123, 194, and 253 kDa within the PM (Derksen and Granados, 1988). In addition, the PMs of virus treated larvae were fragile compared to controls suggesting a physical weakening of the PM structure had occurred. Purified VEF from TnGV was found to significantly increase infectivity of AcMNPV in *T. ni* larvae in a linear dose-dependent manner. The major effect of VEF treatment appeared to be the disruption of the PM, which is the likely basis for increased NPV potency (Gallo et al., 1991).

4.1 Identification and analysis of baculovirus *enhancin* genes

The first gene encoding a VEF was identified in TnGV and sequenced (Hashimoto et al., 1991). Subsequent studies identified *enhancin* genes in several GVs and a few NPVs as listed in Table

1. *Enhancin* genes have been identified in approximately 30% of NPVs and GVs sequenced to date. These genes are more common in GVs, present in about 46% of the genomes analyzed to date vs. approximately 24% of NPVs (Table 1). Several GVs and NPVs contain multiple *enhancin* genes, and the XecnGV contains the most at four. In addition to baculoviruses, *enhancin* genes have been identified in the genomes of microorganisms including *Bacillus cereus* (Ivanova et al., 2003), *Bacillus anthracis* (Read et al., 2003), *Bacillus thuringiensis* (accession no. NC 005957), *Yersinia pestis* (Parkhill et al., 2001), *Salmonella enterica* subsp. enterica serovar Javiana strain GA_MM04042433 (accession no. NZ_ABEH02000001, *Clostridium perfringens* D strain JGS1721 (accession no. ZP_02954459), *Aspergillus oryzae* RIB40 (accession no. XM_001817293), *Enterobacter aerogenes* KCTC 2190 (accession no. NC_015663), and *Listeria ivanovii* subspecies ivanovii PAM 55 (accession no. NC_016011).

Virus	Host	# Enhancin Genes Present	Accession Number[c]
AcMNPV	*Autographa californica*	_	NC_001623
AdhoNPV	*Adoxophyes honmai*	_	NC_004690
AdorGV	*Adoxophyes orana*	_	NC_005038
AgipMNPV	*Agrotis ipsilon*	1	NC_011345
AgseGV	*Agrotis segetum*	1	NC_005839
AgseNPV	*Agrotis segetum*	3	NC_007921
AgMNPV	*Anticarsia gemmatalis*	_	NC_008520
AnpeNPV	*Antheracea pernyi*	_	NC_008035
BmNPV	*Bombyx mori*	_	NC_001962
ChchNPV	*Chrysodeixis chalcites*	_	NC_007151
ChocGV	*Choristoneura occidentalis*	_	NC_008168
CfMNPV	*Choristoneura fumiferana*	1	NC_004778
CfDefNPV	*Choristoneura fumiferana*	_	NC_005137
CfGV[a]	*Choristoneura fumiferana*	1	AF319939
ClbiNPV	*Clanis bilineata*	_	NC_008293
CpGV	*Cydia pomonella*	_	NC_002816
CuniNPV	*Culex nigripalpus*	_	NC_003084
CrleGV	*Cryptophlebia leucotreta*	_	AY_009987
EcobNPV	*Ecotroplis obliqua*	_	NC_008586
EppoNPV	*Epiphyas postvittana*	_	NC_003083
EupsNPV	*Euproctis pseudoconspersa*	1	NC_012693
HearGV	*Helicoverpa armigera*	4	NC_010240
HearMNPV	*Helicoverpa armigera*	1	NC_011615
HearSNPV	*Helicoverpa armigera*	_	NC_002654
HycuNPV	*Hyphantria cunea*	_	NC_007767
HzSNPV	*Helicoverpa zea*	_	NC_003349
LdMNPV	*Lymantria dispar*	2	NC_001973
LyxyMNPV	*Lymantria xylina*	2	NC_013953
LeseNPV	*Leucanua separata*	_	NC_008348
MacoNPV-A	*Mamestra configurata*	1	NC_003529
MacoNPV-B	*Mamestra configurata*	1	NC_004117
MaviNPV	*Maruca vitrata*	_	NC_008725
NeabNPV	*Neodiprion abietis*	_	NC_008252

NeleNPV	*Neodiprion lecontei*	__	NC_005906
NeseNPV	*Neodiprion sertifer*	__	NC_005905
OpMNPV	*Orgyia pseudotsugata*	__	NC_001875
OrleNPV	*Orgyia leucostigma*	__	NC_010276
PhopGV	*Phthorimaea operculella*	__	NC_004062
PlxyMNPV-CL3	*Plutella xylostella*	__	DQ_457003
PlxyGV	*Plutella xylostella*	__	NC_002593
PsunGV	*Pseudaletia unipuncta*	3[a]	NC_013772
RoMNPV	*Rachiplusia ou*	__	NC_004323
SeMNPV	*Spodoptera exigua*	__	NC_002169
SfMNPV	*Spodoptera frugiperda*	__	NC_009011
SpltGV	*Spodoptera litura*	__	NC_009503
SpltMNPV	*Spodoptera litura*	__	NC_003102
TnSNPV	*Trichoplusia ni*	__	NC_007383
TnGV	*Trichoplusia ni*	1[b]	D12617
XcGV	*Xestia c-nigrum*	4	NC_002331

The genus and species of the host organism from which the virus was isolated were used to name the virus using either the first letter of the host genus and species, or the first two letters. NPV stands for nucleopolyhedrovirus, GV for granulovirus, M for multiply enveloped ODV, and S for singly enveloped ODV.

[a] PsunGV does not contain a homologue of HearGV VEF-2 and XecnGV VEF-2 genes. The sequence of the PsunGV VEF-1 sequenced by Roelvink et al., (1995) matches the sequence of Psun VEF-3 reported in the genomic sequence.

[b] The sequence listed for is for only the *enhancin* gene, the genome of this virus has not been sequenced.

[c] References for the accession numbers are AcMNPV, Ayres et al., 1994; AdhoNPV, Nakai et al., 2003; AdorGV, Wormleaton et al., 2003; AgipMNPV, Harrison, 2009; AgseMNPV, Jakubowska et al., 2006; AgMNPV, Oliveira et al., 2006; AnpeMNPV, Fan et al., 2007; BmNPV, Gomi et al., 1999; ChchNPV, van Oers et al., 2005; ChocGV, Escasa et al., 2006; CfMNV, de Jong et al., 2005; CfDefNPV, Lauzon et al., 2005; ClbiNPV, Zhu, S., et al., 2009; CpGV, Luque et al., 2001; CuniNPV, Afonso et al., 2001; CrleGV, Lange & Jehle, 2003; EcobNPV, Ma et al., 2007; EppoNPV, Hyink et al., 2002; EupsNPV, Tang et al., 2009; HearGV, Harrison & Popham, 2008; HearSNPV G4, Chen et al., 2001; HycuNPV, Ikeda et al., 2006; HzSMPV, Chen et al., 2002; LdMNPV, Kuzio et al., 1999; LyxyMNPV, Nai et al., 2010; LeseNPV, Xiao & Qi, 2007; MacoNPV-A, Li, Q., etal., 2002; MacoNPV-B, Li, L., etal., 2002; MaviNPV, Chen et al., 2008; NeabNPV, Duffy, et al., 2006; NeleNPV, Lauzon, et al., 2004; NeseNPV, Garcia-Maruniak, et al., 2004; OpMNPV, Ahrens, et al., 1997; PlxyGV, Hashimoto, et al., 2000; PlxyMNPV-CL3, Harrison & Lynn, 2007; RoMNPV, Harrison & Bonning, 2003; SeMNPV, Ijkel et al., 1999; SfMNPV, Harrison et al., 2008; SpltGV, Wang et al., 2008; SpltMNPV, Pang et al., 2001; TnSNPV, Willis et al., 2005; XcGV, Goto et al., 1998. Sequences only submitted to GenBank: AgseGV, CfGV, HearMNPV, OrleNPV, PhopGV, PsunGV.

Table 1. Sequenced Baculovirus Genomes and Genbank Accession Numbers

Comparison of the LdMNPV-VEF-1 enhancin amino acid sequence with sequences in the BLOCKS database (version 9.0, December 1995 [Henikoff & Henikoff, 1991]) revealed the presence of a signature pattern characteristic of a zinc-binding domain found within metalloproteases (Jongeneel et al., 1989; Murphy et al., 1991). The signature pattern, HEXXH, is sufficient to group a protein into the metalloprotease superfamily. Most baculovirus enhancins have this conserved metalloprotease zinc-binding domain (residues 241 to 246 for the LdMNPV-VEF-1) (Table 2). For this type of enzyme, the zinc ion is chelated by the two histidine residues in this sequence and by a third residue, typically a histidine, cysteine, or aspartic or glutamic acid residue, located anywhere from 20 to 120 aa

LdMNPV-VEF-1	VGAAYYDRNWTAQTNVSMS-RYLQPRATNWLVL	HEIGH	AYDFQFVS--NTPALNEVWNNV 266
LyxyMNPV-VEF-1	VGVAYYGREWTAQTNVSMS-RYLQPRATNWLVL	HEIGH	AYDFQFVI--NTPSLIEVWNNV 266
LyxyNPV-VEF-2	AGIAFYGQHWLGASTNTLL-RYLNVDADPWLIL	HEIGH	GHEFAFVG--TAPPLSEVWTNI 268
LdMNPV-VEF-2	AGIAFYGQHWIGASANTLL-RYLNVDADPWLIL	HEIGH	GHEFAFVG--TAPPLIEVWTNI 268
EupsNPV	SGIAYYGTFWIAHSSYSIS-EYLKADPTNWLAL	HEVGH	AYDFNFVT--NVPVLSEVWTNV 276
XecnGV-VEF-4	HGAAYYSSKWLAMSQSDLS-FFLVPSYTNWLVL	QKLGD	AYNFGFTR--EHTYMAGAWSGM 259
HearGV-VEF-4	HGAAYYSSNWLAMSQSDLS-YFLVPSYTNWLVL	QTLGD	AYNFGFTR--EHTYWAGAWSGM 259
PsunGV-VEF-4	RDTAYYSSNWLAMSQSDLG-KFLEPSFTNWLVL	QTLGD	AYNYGFTR--YDTYWLGVWSGM 259
HearGV-VEF-3	PGGAYYGAFWTAPASTNLG-EYLRVSPTNWMVI	HELGH	AYDFVFT---VNTRLIEIWNNS 267
XecnGV-VEF-3	PGGAYYGPFWTAPANTSLR-DYLVVSPTNWMVI	HELGH	AYDFVFT---VNTRLIEIWNNS 267
PsunGV-VEF-3	PGGAYYGPFWTAPASSNLG-DYLRISPTNWMVI	HELGH	AYDFVFT---VNTILIEIWNNS 267
CfGV	PGGAYYGPFWTAPASSNLG-DYLRISPTNWMVI	HELGH	AYDFVFT---VNTILIEIWNNS 267
TnGV	PGGAYYGPFWTAPASSNLG-DYLRISPTNWMVI	HELGH	AYDFVFT---VNTILIEIWNNS 267
XecnGV-VEF-2	PGAAYYGGSWTANSQSVLG-NYLQVRPGNWLVF	HEIGH	AYDLVFT---QGTNLHEVWNNV 270
HearGV-VEF-2	PGAAYYGGSWTANSQSVLG-NYLQVRPGNWLVF	HEIGH	AYDLVFT---QGTNLHEVWNNV 270
XecnGV-VEF-1	PAEHYYTDAYIANSADTLG-PFLLSTITNWPAL	HQIGH	GYDFNFT---NHTVLIEVWSSV 251
HearGV-VEF-1	PAEHYYTDAYIANSADTLG-PFLLSTITNWPAL	HQIGH	GYDLNFT---NHTVLIEVWSSV 251
PsunGV-VEF-1	PGDHYYSDSYIANSANTLG-PFLLSTITNWPAL	HQIGH	GYDHHFT---NNTSLKEVWSSV 252
AgseNPV-VEF-2	PGDAFYHTWYMGESYETMKSFYLTPSTLNWGAL	HEMAH	SFDIYFARNTIQVSLQEVWTNI 258
AgipMNPV	GGGAFYGHYFMGASSQSMARFFLNLTPLNWGGL	HEIGH	SFDLVFTRNTSQLDVQEVWTNV 256
AgseNPV-VEF-1	PGGAYYSRLYMGESSASMRSFYLRPTPLNWGCL	HEISH	SLDIYFRHNSEQVSLVEVWTNI 254
MacoNPV-B	VGGAYYGKYTMAESSPSMRRFYLTPSKFNWGCL	HEIAH	SFDAHFTSNYIHADIREVWTNI 254
HearMNPV	VGGAYYGKYTMAESSPSMRRFYLTPSKFNWGCL	HEIAH	SFDAHFTSNYIHADIREVWTNI 254
MacoNPV-A	AGGGYYGKYTMGESNPSMRRFYLTPSKFNWGCL	HEIAH	SFDAYFTWNYAHADIREVWTNI 254
AgseGV	SGAAYYGPFWIGVTTQSFD-LFYNVSISNWIML	HVMGH	AYDFEFAN--SKSIFEETWASI 271
CfMNPV	SGAAYYNNLTMGQSNSSVEHFYLRPLPTNWGGL	HEIAH	AYDFHFVRS-GPVPLNEVWNNI 265

LdMNPV-VEF-1	LADRYQYDFMSFDERQRDASVYENGNRDRVERNIAERIDNRAPY--ESWSFFQKMAVFTW 324
LyxyMNPV-VEF-1	LADRYQYDFMSFDERQRDASVYENGNRDRVERNIAELIRNRVPY--ENWSFFQKLTVFTW 324
LyxyNPV-VEF-2	LPNLFQFDTMTSAERETAAWIYDYGRREAVERGLGALIDQRVDY--DKFSFRERLFAYAP 326
LdMNPV-VEF-2	LPNLFQFDTMTSAERQTAAWIYDHGRREVVERGLGALIDQRVDY--DKFSFRERLFAYAP 326
EupsNPV	LPDRYQFYNMTYEERQRLSWIFGEGNRQTVEKSLQSLISSKTKYSSDDYYFRERLLTLGL 336
XecnGV-VEF-4	LSDRLQYYHQTRAERQTVSEIYAG-QRLRVEEEIAAVLLDNLNL--DLWTELHKIVFFTW 316
HearGV-VEF-4	LSDRLQYYHQTRAERQTISEVYAG-QRLRVEEEIAAVLLDNLNL--DLWTELHKIVFFTW 316
PsunGV-VEF-4	LSDRLQYYYQDISERQTLSKIYAG-ERNRIEEEINVTIQDDLNL--DLWTELHKIVFFTW 316
HearGV-VEF-3	FCDRIQYTWMNKTKRQQLARIYEN-QRPQKEAAIQALIDNNVPF--DNWDFFEKLSIFAW 324
XecnGV-VEF-3	FCDRIQYTWMNKTKRQQLARIYEN-QRPQKEAAIQALIDNNVPF--DNWDFFEKLSIFAW 324
PsunGV-VEF-3	LCDRIQYKWMNKTKRQQLARVYEN-RRPQKEATIQALIDNNSPF--DNWGFFERLIIFTW 324
CfGV	LCDRIQYKWMNKTKRQQLARVYEN-RRPQKEATIQALIDNNSPF--DNWGFFERLIIFTW 324
TnGV	LCDRIQYKWMNKIKRQQLARVYEN-RRPQKEATIQALIDNNSPF--DNWGFFERLIIFTW 324
XecnGV-VEF-2	YGDRMQYQLMDAAVRQSSASVYENGNRVRVENNIMNLIDTNVNY--SSWSFFQKLAMFTF 328
HearGV-VEF-2	YGDRMQYQLMDAAVRQSSASVYENGNRVRVENNIMNLIDTNVNY--SSWSFFQKLAMFTF 328
XecnGV-VEF-1	LADSMQYQWMNTAERQKLATIYDGDQREYAETEIVKNLAESASF--NNLDAQWRTILLSF 309
HearGV-VEF-1	LADSMQYQWMNTAERQKLPTIYDGDQREYAETEIVKNLAESASF--NNLDAQWRTILLAF 309
PsunGV-VEF-1	LADSMQYQWMNTTERQNLPMIYDGDQREHAEAEIVKELAESVAF--HDLDTNWRTILLAF 310
AgseNPV-VEF-2	WPDYYQYTRLTTDEYEDKSWMLSP-TREVSIGQLVEKFHTTTLH---DWDHRERLLYLTA 314
AgipMNPV	LPDYYQHTRLSPDEYEANAWLLDP-GREATFGQLIAKFHTTPIH---EWNLRERLIFLTS 312
AgseNPV-VEF-1	FPDFYQYSKLTPAEYQASAWIMGS-SQQTIWGNLIAKFHTVSVH---DWDLRERLIYLVQ 310
MacoNPV-B	MPDYYQYLNYTEEEYLTRGWKYDG-RRDAVLIEIKRIFDVIPFN---KWSLRQRLVFLTS 310
HearMNPV	MPDYYQYLNYTEEEYLTRGWKYDG-RRDAVLIEIKRIFDVIPFN---KWSLRQRLVFLTS 310
MacoNPV-A	MPDYYQYLNFTEEEYLTKSWKLDG-QRDTLFMEIKALFGVVPFN---EWYLRDRLVFLTS 310
AgseGV	FADRYQFFRSTKMERYLTNTVFGPQMYSQIITDINILFSARTPFP--AWPEFRKLLILSL 329
CfMNPV	LCDAYQSRYLELDEKS-----VNSPNSLTMGRNIVRKIENGKVF--DSYNLFEKLAVFSY 318

[a] The conserved HEXXH sequence is highlighted in turquoise, the non-conserved corresponding sequence in a few enhancins is highlighted in yellow, and aspartic and glutamic residues 20 or more amino acids downstream are highlighted in red and green, respectively.

Table 2. Alignment of Baculovirus Enhancin Proteins with COBALT (NCBI)[a] downstream of the HEXXH sequence (Häse & Finkelstein, 1993; Jiang & Bond, 1992). There are several aspartic and glutamic acid residues between 20 and 120 aa from the HEXXH sequence that are present in baculovirus enhancins, any of which could function as a third zinc-binding ligand in the enhancin proteins (Table 2). In the metalloproteases, the glutamic acid residue within the HEXXH sequence is the catalytic base, which polarizes a water molecule involved in the nucleophilic attack of the peptide bond to be cleaved. XecnGV-VEF-1, HearGV-VEF-1, PsunGV-VEF-1 contain a glutamine residue in place of the glutamic acid within the HEXXH (HQXXH) consensus, in XecnGV-VEF-4, HearGV-VEF-4, PsunGV-VEF-4 a QXXDG sequence is in place of the consensus sequence, and AgseGV contains a HVMGH sequence in place of the consensus sequence (Table 2). The alterations in the zinc binding domain of these enhancins could have made these proteins non-functional. If so, a virus with a mutated non-functional *enhancin* gene could gain selective advantage through acquisition of a functional *enhancin* gene. These events offer a basis for the presence of multiple *enhancin* genes within viral genomes. A functional analysis of the enhancins in XecnGV-VEF-1, HearGV-VEF-1, PsunGV-VEF-1, XecnGV-VEF-4, HearGV-VEF-4, PsunGV-VEF-4, and AgseGV has not been performed.

A comparison of the number of amino acids in baculovirus enhancins and the proteins that have the most and least amino acid identities are listed in Table 3. Baculovirus enhancins exhibit a great deal of heterogeneity. For example, LdNPV-VEF-1 is 89% identical to LyxyMNPV-VEF-1. In contrast, LdNPV-VEF-1 is only 16% identical to MacoNPV-A (Table 3). The size of baculovirus enhancins is from 758 amino acids for CfMNPV to 1004 amino acids for AgseGV.

A phylogenetic analysis of all currently known baculoviruses was performed using the phylogenetic tree function of CLUSTAL-W, and the tree is shown in Figure 1. All of the GV enhancins, with the exception of AgseGV, form a group, all of the NPV enhancins form a group, and AgseGV is within its own group (Fig 1). Three subgroups are within the GV group; XecnGV-VEF-2 - PsunGV-VEF-1, HearGV-VEF-3 - TnGV, and XecnGV - Psun-VEF-4. Two primary groups are present within the NPV enhancin group; LdMNPV-VEF-1 - LdMNPV-VEF-2, and CfMNPV - MacoNPV-A. The high level of heterogeneity exhibited by the baculovirus enhancins may suggest that these genes arose in viral genomes from independent sources.

The presence of *enhancin* genes in bacteria suggests a possible means for enhancin gene exchange between microorganisms. The *B. cereus* group contains the closely related organisms *B. cereus*, an opportunistic pathogen of humans; *B. anthracis*, a mammalian pathogen; and *B. thuringiensis*, an insect pathogen. The presence of *enhancin* genes in *B. cereus* and *B. anthracis* led to the suggestion that these organisms evolved from an ancestor of the *B. cereus* group that was an opportunistic insect pathogen (Ivanova et al., 2003; Read et al., 2003). The subsequent finding of an *enhancin* gene in *B. thuringiensis* provides further support for this hypothesis. If the *B. cereus* ancestor resided in the guts of insects that were NPV hosts, there may have been an opportunity for an exchange of genetic material between these bacteria and NPVs. The enhancin in *Y. pestis*, the causative agent of the disease referred to as plague, may aid its colonization of the flea. The *Y. pestis enhancin* gene is flanked by a tRNA gene and transposase fragments, which may suggest that this bacteria obtained its *enhancin* gene via horizontal transfer. However, a recent study found that expression of bacterial enhancins in insect cells caused cytotoxicity, and they did not enhance viral infectivity (Galloway et al., 2005). These results may suggest that bacterial and baculovirus enhancins have evolved different functions.

Viral Enhancin	# of Amino Acids	Viral Enhancin with the Greatest Amino Acid Identity, % Identity	Viral Enhancin with the Least Amino Acid Identity, % Identity
LdMNPV-VEF-1	782	LyxyMNPV-VEF-1, 89%	MacoNPV-A, 16%
LyxyMNPV-VEF-1	782	LdMNPV-VEF-1, 89%	AgseNPV-VRF-2, 17%
EupsNPV	802	LyxyMNPV-VEF-2, 34%	MacoNPV-A, 13%
LyxyMNPV-VEF-2	788	LdMNPV-VEF-2, 94%	MacoNPV-A, 12%
LdMNPV-VEF-2	788	LyxyMNPV-VEF-2, 94%	MacoNPV-B, 12%
XecnGV-VEF-2	867	HearGV-VEF-2, 96%	MacoNPV-B, HearMNPV, AgipMNPV, 13%
HearGV-VEF-2	865	XecnGV-VEF-2, 96%	AgseNPV-VEF-2 AgipMNPV, 13%
HearGV-VEF-3	902	XecnGV-VEF-3, 86%	HearMNPV, 15%
PsunGV-VEF-3	901	TnGV, CfGV, 98%	AgseNPV-VEF-2, 12%
TnGV	901	PsunGV-VEF-3, CfGV, 98%	AgseNPV-VEF-2, 12%
CfGV	901	PsunGV-VEF-3, TnGV, 98%	AgipMNPV, 14%
XecnGV-VEF-3	898	HearGV-VEF-3, 86%	AgipMNPV, 15%
XecnGV-VEF-4	856	HearGV-VEF-4, 95%	AgseNPV-VEF-2, AgipMNPV, 11%
HearGV-VEF-4	856	XecnGV-VEF-4, 95%	AgseNPV-VEF-2, 11%
PsunGV-VEF-4	857	XecnGV-VEF-1, 23%	AgseNPV-VEF-1, 11%
XecnGV-VEF-1	824	HearGV-VEF-1, 96%	AgipMNPV, 10%
HearGV-VEF-1	823	Xecn-VEF-1, 98%	AgipMNPV, 10%
CfMNPV	758	AgseNPV-VEF-2, 19%	PsunGV-VEF-1, 16%
PsunGV-VEF-1	828	AgseGV, 21%	AgipMNPV, 12%
AgseGV	1004	PsunGV, 21%	HearMNPV, 13%
AgseNPV-VEF-1	877	AgseNPV-VEF-2, 43%	PsunGV-VEF-4, 11%
MacoNPV-B	848	HearMNPV, 99%	LdMNPV-VEF-2, 12%
HearMNPV	848	MacoNPV-B, 99%	AgseGV, XecnGV-VEF-2, 13%
AgseNPV-VEF-2	883	AgipMNPV, 53%	Xecn-VEF-4, HearGV-VEF-4, 11%
AgipMNPV	897	AgseNPV-VEF-2, 53%	Xecn-VEF-1, HearGV-VEF-1, 10%
MacoNPV-A	847	HearNPV, 81%	LyxyMNPV-VEF-2, 12%

Table 3. Number of Amino Acids in Baculovirus Enhancins and Identity of the Enhancin Most and Least Similar.

4.2 Function of baculovirus enhancins

Studies on proteins within TnGV granules identified a 98 kDa protein that enhanced AcMNPV potency in bioassays (Gijzen et al., 1995). The enhancin from TnGV granules was purified by gel filtration and ion exchange chromatography and was found to be a metalloprotease based on activity inhibition studies using metal chelators (Lepore et al., 1996). In addition, the *enhancin* gene from TnGV was expressed by a recombinant AcMNPV containing this gene, the enhancin protein was purified, and was found to enhance the infectivity of AcMNPV in larval bioassays, thereby confirming that the *enhancin* gene codes for the enhancing factor in TnGV granules (Lepore et al., 1996).

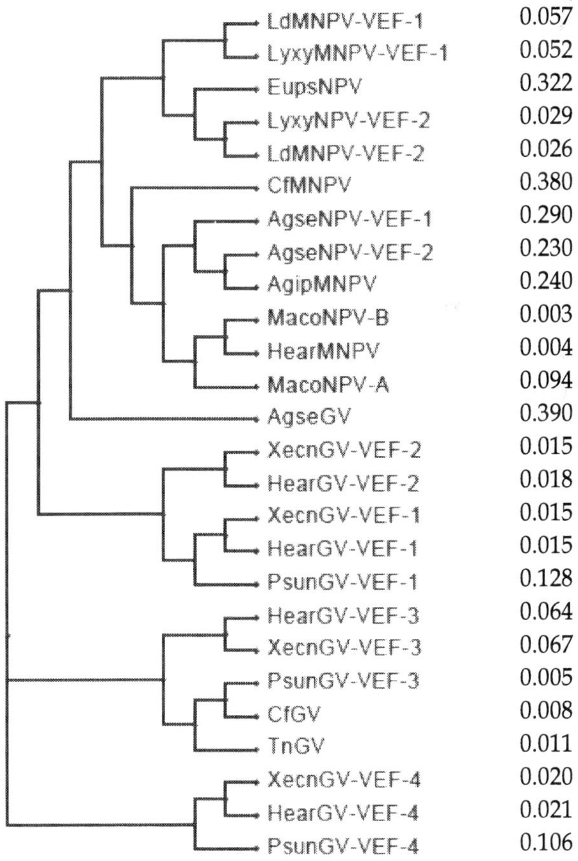

Fig. 1. Phylogenetic tree of NPV and GV enhancins. The distances from the nodes are shown on the right of the figure.

Studies with isolated PMs from *T. ni* and *P. unipuncta* larvae found that PMs treated with purified TnGV enhancin were significantly more permeable to AcMNPV compared to control PMs. These results provide evidence that the PM is a barrier to virus movement across the PM and that TnGV enhancin facilitates infection by altering the permeability of the PM (Peng et al., 1999). Wang and Granados (1997) found that the target substrate for GV enhancins is insect intestinal mucin, a glycosylated protein associated with the PM. Degradation of intestinal mucin increased access of virions to the midgut epithelial cells, increasing susceptibility of the host insect to viral infection (Wang & Granados, 2000). This is likely the basis for the early observations of enhancement of NPV infections *in vivo* when co-administered with GV granules and purified enhancin. Based on these findings, and the finding the enhancins constitute 5% of the protein in GV granules, there is substantial evidence supporting the hypothesis that GV enhancins facilitate GV virus movement through the PM.

However, earlier studies by Tanada and colleagues generated evidence that GV enhancins bind to larval midgut cells at specific sites, and they hypothesized that the enhancin protein serves as a receptor for some NPVs (Tanada, 1985). They found that significantly more virus particles were attached to or already in midgut microvilli in insects treated with enhancin plus PsunNPV compared to insects without enhancin, even though polyhedra and ODV were found in abundance next to microvilli in the control insects. In addition, enhancin was found to be associated with midgut cell microvilli when purified protein was administered *per os* (Tanada et al., 1980), and *P. unipuncta* larval midgut cells were found to contain specific binding sites for GV enhancins that enhance the infection process of NPVs (Uchima *et al.*, 1988). Also, several investigations found that GV enhancins increased infection of NPVs in cell culture systems (Hukuhara & Zhu, 1989; Kozuma & Hukuhara, 1994; Tanada, 1985).

Studies on the function of baculovirus enhancins have been within the context of heterologous systems using GV enhancins/granules to investigate their impact on the potency of NPVs administered *per os*, and the infectivity of NPV BV in cell culture systems. After identification and characterization of the TnGV gene encoding the enhancin protein by Granados and colleagues, studies focused on enhancin biochemical properties and impacts on the PM. However, a direct impact of GV enhancin on GV potency through deletion of the gene from the GV genome has not been demonstrated. Studies on the function of LdMNPV enhancins described below were the first studies performed within a homologous system.

4.3 Analysis of enhancin function(s) in the LdMNPV

The LdMNPV was the first NPV found to contain an *enhancin* gene (Bischoff & Slavicek, 1991), and upon sequencing the viral genome, a second *enhancin* gene was identified (Kuzio et al., 1999). *Enhancin 1* and *2* gene transcripts are expressed at late times after infection from a consensus baculovirus late promoter (Bischoff & Slavicek, 1991; Popham et al., 2001). To investigate the function of the LdMNPV *enhancin 1* gene a recombinant LdMNPV virus was constructed that lacked a functional *enhancin 1* gene (E1cat). Potency analysis through *L. dispar* larval bioassays revealed that the *enhancin 1* gene minus viral strain was approximately 2-3 fold less potent than wild-type (WT) viruses, suggesting that the LdMNPV enhancin affects viral potency (Bischoff & Slavicek, 1997). The effect of the second LdMNPV *enhancin* gene alone and in conjunction with the *enhancin 1* gene on viral potency was investigated through bioassay using two recombinant viruses, one with a deletion in the *enhancin 2* gene (E2del) and a second with deletion mutations in both *enhancin* genes (E1delE2del) (Popham et al., 2001). The *enhancin* gene viral constructs were verified by Southern analysis and were shown not to produce *enhancin* gene transcripts by Northern analysis. The E2del virus exhibited an average decrease in viral potency of 1.8 fold compared to wild-type virus. In the same bioassays, the recombinant virus E1cat, which does not produce an *enhancin1* gene transcript, exhibited an average decrease in viral potency of 2.3 fold compared to control virus. The E1delE2del virus exhibited an average decrease in viral potency of 12 fold compared to wild-type virus, indicating that the two genes confer a non-additive compensatory enhancement of viral efficacy (Popham *et al.*, 2001). Collectively, these results suggest that both LdMNPV *enhancin* genes contribute to viral potency, that each enhancin protein can partially compensate for the lack of the other protein, and that both *enhancin* genes are necessary for wild-type viral potency. Detergent dissociation studies and immuno-electron microscopy were used to localize LdMNPV enhancins in viral structures (Slavicek & Popham, 2005). Polyclonal antibodies specific to LdMNPV enhancin 1 and enhancin 2 identified unique proteins of 84 and 90 kDa, respectively, at 48 h post infection (p.i.) in infected cell extracts prepared from 0 to 96 h p.i. The 84- and 90-kDa proteins are close to the predicted sizes of Enhancin 1 (89.2) and Enhancin 2 (88.4), and their appearance at 48 h p.i. occurred at the same time that *enhancin 1* and 2 gene transcripts were first detected in earlier studies (Bischoff & Slavicek, 1997; Popham et al., 2001). Enhancin 1 and 2 were found in polyhedra and were further localized to ODV. The amounts of Enhancin 1 and 2 isolated from polyhedra and from ODV isolated from the same number of polyhedra were approximately the same, suggesting that most, if not all, Enhancin 1 and 2 present in polyhedra were located within ODV. Enhancin 1 and 2 were found not to be components of BV. Treatment of ODV with detergents indicated that Enhancin 1 and 2 were present in ODV envelopes, and that the Enhancins may have an ionic bond with the nucleocapsid through positively charged amino acids. Immunogold labeling of polyhedron sections localized the Enhancins to ODV envelopes, either at the outside edge or within the envelope. The location of LdNPV enhancins within ODV envelopes (Slavicek and Popham, 2005) would be consistent with the hypothesis of a role in binding to midgut cells. A general PM degradative function of GV enhancin is consistent with their location and amount within the GV granule. However, this does not preclude a duel function of binding to and facilitating entry into midgut cells. Such a duel role for GV enhancin has never been tested within a homologous context. As noted earlier, the presence of LdMNPV enhancins within ODV envelopes provides a position for the proteins to interact directly with midgut cells as well as on the PM as ODV traverses this host barrier.

To determine if the LdMNPV enhancin's sole function was degradation of the PM, a fluorescent brightener was used to eliminate the PM, and the impact on the potency of recombinant virus lacking both *enhancin* genes was determined (Hoover et al., 2010). Removal of the PM should eliminate the reduced potency of the recombinant virus lacking both *enhancin* genes compared to wild-type virus. The potency of the *enhancin* gene double deletion virus was about 14-fold less potent compared to wild-type virus in brightener treated larvae. These results suggest that the LdNPV *enhancin* genes have a function that impacts viral potency that does not involve degradation of the PM. The findings that the two enhancin proteins can partially compensate for the lack of the other (Popham et. al., 2001), their location in the ODV envelope (Slavicek & Popham, 2005), and the reduced potency of *enhancin* gene deletion viruses in the absence of the PM support the hypothesis that one or both of the enhancins are involved in a function that impacts viral potency, such as binding to midgut epithelial cells in addition to disruption of the PM. Further studies are needed to address this possibility.

4.4 Alpha-helix transmembrane domain analysis of NPV and GV enhancins

Analyses of LdMNPV Enhancin 1 and 2 for the presence of transmembrane alpha-helices by the DAS method (http://www.sbc.su.se/_miklos/DAS/), the PRED-TMR Algorithm (http://o2.db.uoa.gr/PRED-TMR), TMHMM (www .cbs.dtu.dk/services/TMHMM/), and SOSUI (http://sosui.proteome.bio.tuat.ac.jp/sosuiframe0.html) all predict the presence of a transmembrane alpha-helix in LdMNPV Enhancin 1 in the amino acid region from about positions 738 to 761 and in Enhancin 2 in the amino acid region from about positions 747 to 771 (Table 4) (Slavicek & Popham, 2005). The TMHMM method further predicts that amino acids of Enhancin 1 from positions 1 to 737 and of Enhancin 2 from positions 1 to 746 are located on the outside of the membrane structure, residues 738 to 760 (Enhancin 1) and 747 to 769 (Enhancin 2) span the membrane, and amino acids 761 to 782 (Enhancin 1) and 770 to 788 (Enhancin 2) are located on the inside of the membrane structure (i.e., next to the nucleocapsid). These predictions suggest that the carboxyl-terminal regions of Enhancin 1 and 2 are anchored within the ODV envelope, and that the majority of the proteins extend beyond the envelope surface. An interesting correlation is that many of the gold particles were found at the outside edge of ODV envelopes after immunogold labeling, which is consistent with the Enhancin 1 and 2 epitopes (regions 279 to 298 and 506 to 525, respectively) being located on the outside of ODV envelopes. If this orientation were present, the LdMNPV enhancins would be positioned to interact with the peritrophic membrane within the host midgut. It is interesting that the regions from positions 761 to 782 of Enhancin 1 and 770 to 788 of Enhancin 2 each contain six basic amino acids, which may bind the enhancins to the nucleocapsids (Table 5).

Analysis of the NPV enhancins with TMHMM, PRED-TMR2, DAS and SOSUI transmembrane protein structure prediction programs predicted the presence of alpha-helix transmembrane domains near the carboxyl terminus in all NPV enhancins (Table 4). In contrast, none of the GV enhancins were predicted to contain transmembrane protein structures, and the programs classified them as soluble proteins. The lack of GV enhancin transmembrane domains and classification of GV enhancins as soluble proteins are consistent with their localization as components of granules. All of the NPV enhancins contain from 2 to 9 basic amino acid residues between the end of the transmembrane region and the carboxyl terminus (Table 5). The CfNPV enhancin contained the least (2) and AgipMNPV contained the most (9) basic residues. All of the basic residues in this region in MacoNPV-A, HearMNPV, and Maco-B were lysine residues and the other NPV enhancins contained arginine and lysine residues, lysine and histidine residues, or all three (Table 5).

	Amino Acids Constituting the Predicted Alpha-Helix Transmembrane Domain by the Respective Prediction Program			
Viral Enhancin	TMHMM	PRED-TMR2	DAS	SOSUI
LdMNPV-VEF-1	738-760	744-761	738-761	738-760
LyxyMNPV-VEF-1	740-762	746-763	739-763	739-761
EupsNPV	761-783	763-779	761-782	762-784
LyxyMNPV-VEF-2	747-769	750-768	748-771	749-771
LdMNPV-VEF-2	747-769	750-768	748-771	749-771
CfMNPV	719-741	720-740	718-741	719-741
AgseNPV-VEF-1	819-841	822-841	816-843	817-839
MacoNPV-B	806-828	805-824	803-823	803-825
HearMNPV	806-828	805-824	803-827	803-825
AgseNPV-VEF-2	827-849	825-844	822-849	827-849
AgipMNPV	827-849	827-846	823-852	828-850
MacoNPV-A	804-826	804-823	803-826	803-825
XecnGV-VEF-2	---	---	---	---
HearGV-VEF-2	---	---	---	---
HearGV-VEF-3	---	---	---	---
PsunGV-VEF-3	---	---	---	---
TnGV	---	---	---	---
CfGV	---	---	---	---
XecnGV-VEF-3	---	---	---	---
XecnGV-VEF-4	---	---	---	---
HearGV-VEF-4	---	---	---	---
PsunGV-VEF-4	---	---	---	---
XecnGV-VEF-1	---	---	---	---
HearGV-VEF-1	---	---	---	---
PsunGV-VEF-1	---	---	---	---
AgseGV	---	---	---	---

Table 4. Analysis of NPV and GV Enhancins for Transmembrane Alpha-helices and Carboxyl Terminal Basic Amino Acids

Viral Enhancin	Amino Acid End of Transmembrane Domain	Downstream Sequence[a]
LdMNPV-VEF-1	761	VNRRGRQSPKAAERAPPLQRV
LyxyMNPV-VEF-1	763	VNQRGRQNPKAAEHAPLQHS
EupsNPV	779	KLVYSKTTNITESTPLMLDRQQT
LyxyMNPV-VEF-2	768	IATIARRAKRDDARPPSSIKA
LdMNPV-VEF-2	768	TIARRAKRDDARPPSVIKA
CfMNPV	740	KNMATPNTSHNLAPNIS
AgseNPV-VEF-1	841	LKVTRAQETAPLTPIPAISAPIASAPTQTRRRRK IIE
MacoNPV-B	824	PNAIETIIKEKPKTNIKSIK
HearMNPV	824	IKIASPSKKQVITKEKPKPVIKSIK
AgseNPV-VEF-2	844	VKFLVGANKCAIAETIQAPPPPGKTITTRRPTR VTPITTA
AgipMNPV	846	IKLIVSPNCVVYQTSPSLPPPPAASARTRTARN VTSRRPTRGVINRSPTPTR
MacoNPV-A	823	SPSKKQVITKEKPKPVIKSIK

a. Arginine residues are highlighted in turquoise, histidine residues in green, and lysine residues in yellow.

Table 5. Location of Basic Amino Acids Downstream of the Alpha-helix Transmembrane Domain

4.5 Applications of baculovirus enhancins for insect pest control

Researchers have been working to develop means of using baculovirus *enhancin* genes to increase the efficacy of insect control efforts. The impact on potency of AcMNPV by TnGV enhancin was indirectly investigated by combining cell culture cells infected with a recombinant virus containing the TnGV *enhancin* gene. The potency of AcMNPV and SeNPV were 21-fold and 10-fold greater, respectively when the cell culture cells infected with the enhancin expressing AcNPV recombinant were present (Hayakawa et al. 2000). More recently, a recombinant AcMNPV was generated that expressed the TnGV *enhancin* gene. Polyhedra from the enhancin expressing virus were found to be approximately 2-fold more potent compared to wild-type AcMNPV polyhedra (Del Rincon-Castro & Ibarra, 2005). Insertion of the MacoA *enhancin* gene into AcMNPV increased the potency of the recombinant virus approximately 4-fold (Qianjun et al., 2003). The TnGV *enhancin* gene was also placed into the genome of tobacco and found to impact viral potency. Larvae feeding on diet containing dried tobacco leaves expressing TnGV enhancin exhibited a 10-fold enhancement of AcMNPV infection (Hayakawa et al., 2000). TnGV enhancin was found to increase the toxicity of *B. thuringiensis* (Berliner) in larval bioassays with *T. ni, H. zea, H. virescens, S. exigua, P. includes,* and *A. gemmatalis* (Granados et al., 2001).

5. Conclusions

The location of NPV enhancins within ODV envelopes is likely a conserved characteristic of all NPV enhancins, in contrast to GV enhancins being located within granules. This difference suggests that the NPVs and GVs utilize distinct approaches to degradation of the PM. Results to date indicate that GVs release a large amount of enhancin into the larval midgut, which then degrades the PM in a non-targeted random manner. In contrast, the NPVs may utilize enhancins located in ODV envelopes to "tunnel" through the PM to gain access to larval midgut cells. The presence of transmembrane domains in the carboxyl terminus of all NPV enhancins known to date suggests that the localization of LdMNPV enhancins to ODV envelopes is a conserved feature of NPV enhancins. Whether this is a conserved characteristic will require analyses of the locations of enhancins in additional NPVs. The finding that deletion of LdMNPV enhancins negatively impacts viral potency in the absence of the PM supports the hypothesis that the LdMNPV enhancins have a function that impacts viral potency that is distinct from degradation of the PM. This could support earlier hypothesis suggesting that GV enhancins increase NPV potency in heterogonous systems through ODV binding to midgut epithelial cells. The use of NPV and GV enhancins to increase the potency of NPVs lacking enhancins and impact on NPV potency when expressed in plants indicates that enhancins can be used to increase insecticidal activity of NPVs and may be useful for commercial pest control applications.

6. References

Ackermann, H-W. & Smirnoff, W.A. (1983). A morphological investigation of 23 baculoviruses. *J. Invertebr. Pathol.* 41:269-280.

Afonso, C.L., Tulman, E.R., Lu, Z., Balinsky, C.A., Moser, B.A., Becnel, J.J., Rock, D.L. & Kutish, G.F. (2001). Genome sequence of a baculovirus pathogenic for *Culex nigripalpus*. *J. Virol.* 75:11157–11165.

Ahrens, C.H., Russell, R., Funk, C.J., Evans, J.T., Harwood, S.H. & Rohrmann, G.F. (1997). The sequence of *Orgyia pseudotsugata* multinucleocapsid nuclear polyhedrosis virus genome. *Virology* 229:381–399.

Ayres, M. D., Howard, S. C., Kuzio, J., Lo´ pez-Ferber, M. & Possee, R. D. (1994). The complete DNA sequence of *Autographa californica* nuclear polyhedrosis virus. *Virology* 202:586–605.

Bischoff, D. S., & Slavicek, J.M. (1997). Molecular analysis of an *enhancin* gene in the *Lymantria dispar* Nuclear Polyhedrosis Virus. *J. Virol.* 71:8133-8140.

Black, B.C., Brennan, L.A., Dierks, P.M. & Gard, I.E. (1997). Commercialization of baculovirus pesticides. In *The Baculoviruses* (L.K. Miller, Ed.), pp. 341-387. Plenum, New York.

Bonning, B.C. & Hammock, B.D. (1996). Development of recombinant baculoviruses for insect control. *Annu. Rev. Entomol.* 41:191-210.

Chen, Y-R., Wu, C-Y., Lee, S-T., Wu, Y-J., Lo, C-F., Tsai, M-F. & Wang, C-H. (2008). Genomic and host range studies of *Maruca vitrata* nucleopolyhedrovirus. *J. Gen. Virol.* 89:2315-2330.

Chen X.W., IJkel, W.F.J., Tarchini, R., Sun, X.L., Sandbrink, H., Wang, H.L., Peters, S., Zuidema, D., Lankhorst, R.K., Vlak, J.M. & Hu, Z.H. (2001). The sequence of the

Helicoverpa armigera single nucleocapsid nucleopolyhedrovirus genome. *J. Gen. Virol.* 82:241–257.

Chen, X.W., Zhang, W.J., Wong, J., Chun, G., Lu, A., McCutchen, B.F., Presnail, J.K., Herrmann, R., Dolan, M., Tingey, S., Hu, Z.H. & Vlak, J.M. (2002). Comparative analysis of the complete genome sequences of *Helicoverpa zea* and *Helicoverpa armigera* single-nucleocapsid nucleopolyhedroviruses. *J. Gen. Virol.* 83: 673–684.

Crouch, E.A., Cox, L.T., Morales, K.G. & Passarelli, A.L. (2007). Inter-subunit interactions of the *Autographa californica* M nucleopolyhedrovirus RNA polymerase. *Virology* 367: 265–274.

d'Alencon, E., Piffanelli, P., Volkoff, A.N., Sabau, X., Gimenez, S., Rocher, J., Cerutti, P. & Fournier, P. (2004). A genomic BAC library and a new BAC-GFP vector to study the holocentric pest *Spodoptera frugiperda*. *Insect Biochem. Mol. Biol.* 34:331–41.

de Jong, J.G., Lauzon, H.A.M., Dominy, A.P., Carstens, E.B., Arif, B.M. & Krell, P.J. (2005). Analysis of the *Choristoneura fumiferana* nucleopolyhedrovirus genome. *J. Gen. Virol.* 86:929-943.

Del Rincon-Castro, M.C. & Ibarra, J.E. (2005). Effect of a nucleopolyhedrovirus of Autographa californica expressing the *enhancin* gene of *Trichoplusia ni* granulovirus on *T. ni* larvae. *Biocon. Sci. & Technol.* 15:701-710.

Derksen, A. C. G. & Granados, R. R. (1988). Alteration of a lepidopteran peritrophic membrane by baculoviruses and enhancement of viral infectivity. *Virology* 167:242-250.

Duffy, S.P., Young, A.M., Morin, B., Lucarotti, C.J., Koop, B.F. & Levin, D.B. (2006). Sequence analysis and organization of the *Neodiprion abietis* nucleopolyhedrovirus genome. *J. Virol.* 80:6952-6963.

Engelhard, E. K., Kammorgan, L. N. W., Washburn, J. O. & Volkman, L. E. (1994). The insect tracheal system: a conduit for the systemic spread of *Autographa californica* M nuclear polyhedrosis virus. *Proc. Natl. Acad. Sci. USA* 91:3224–3227.

Erlandson, M. (2008). Insect pest control by viruses. In: *Encyclopedia of Virology*, Third Edition, 3:125-133.

Escasa, S.R., Lauzon, H.A.M., Mathur, A.C., Krell, P.J. & Arif, B.M. (2006). Sequence analysis of the *Choristoneura occidentalis* granulovirus genome. *J. Gen. Virol.* 87:1917-1933.

Fan, Q., Li, S., Wang, L., Zhang, B., Ye, B., Zhao, Z., & Cui, L. (2007). The genome sequence of the multinucleocapdis nucleopolyhedrovirus of the Chinese oak silkworm *Antheraea pernyi*. *Virology* 366:304-301.

Federici, B. A. (1997). Baculovirus pathogenesis, p. 33-59. *In* L. K. Miller (ed.), *The Baculoviruses*. Plenum Press, New York, N.Y.

Fang, M., Nie, Y., Wang, Q., Deng, F., Wang, R., Wang, H., Wang, H., Vlak, J.M., Chen, X. & Zu, Z. (2006). *Open reading frame 132 of Helicoverpa armigera nucleopolyhedrovirus encodes a functional per os infectivity factor (PIF-2)*. *J. Gen. Virol.* 87:2563–2569.

Fang, M., Nie, Y., Harris, S., Erlandson, M.A. & Theilmann, D.A. (2009). *Autographa californica* multiple nucleopolyhedrovirus core gene ac96 encodes a per os infectivity factor (PIF-4). *J. Virol.* 83:12569–12578.

Faulkner, P., Kuzio, J., Williams, G.V. & Wilson, J.A. (1997). *Analysis of p74, a PDV envelope protein of Autographa californica* nucleopolyhedrovirus required for occlusion body infectivity in vivo. *J. Gen. Virol.* 78:3091–3100.

Flipsen, J.T., Martens, J.W., van Oers, M.M., Vlak, J.M. & van Lent, J.W. (1995). *Passage of Autographa californica* nuclear polyhedrosis virus through the midgut epithelium of *Spodoptera exigua* larvae. *Virology* 208:328–35.

Gallo, L. G., Corsaro, B.G., Hughes, P.R., & Granados, R.R. (1991). In vivo enhancement of baculovirus infection by the viral enhancing factor of a granulosis virus of the cabbage looper, *Trichoplusia ni* (Lepidoptera: Noctuidae). *J. Invertebr. Pathol.* 58:203-210.

Galloway, C.S., Wang, P., Winstanley, D., & Jones, I.M. (2005). Comparison of the bacterial Enhanin-like proteins from *Yersinia* and *Bacillus* spp. with a baculovirus Enhancin. *J. Invertebr. Pathol.* 90:134-137.

Garcia-Maruniak, A., Maruniak, J.E., Zanotto, P.M., Doumbouya, A.E., Liu, J.C., Merritt, T.M. & Lanoie, J.S. (2004). Sequence analysis of the genome of the *Neodiprion sertifer* nucleopolyhedrovirus. *J. Virol.* 78:7036–7051.

Gijzen, M., Roelvink, P. & Granados, R. (1995). Characterization of viral enhancing activity from *Trichoplusia ni* granulosis virus. *J. Invertebr. Pathology* 65:289-294.

Gomi, S., Majima, K. & Maeda, S. (1999). Sequence analysis of the genome of *Bombyx mori* nucleopolyhedrovirus. *J. Gen. Virol.* 80:1323-1337.

Goto, C., Hayakawa, T. & Maeda, S. (1998). Genome organization of *Xestia c-nigrum* granulovirus. *Virus Genes* 16:199-210.

Goto, C. 1990. Enhancement of a nuclear polyhedrosis virus (NPV) infection by a granulosis virus (GV) isolated from the spotted cutworm, *Xestia c-nigrum* L. (Lepidoptera: Noctuidae). *Appl. Entom. Zool.* 25:137-137.

Granados, R.R. 1980. Infectivity and mode of action of baculoviruses. *Biotechnol. Bioeng.* 22:1377-1405.

Granados, R.R. & Lawler, K.A. (1981). In vivo pathway of *Autographa californica* baculovirus invasion and infection. *Virology* 108:297-308.

Granados, R.R., Fu, Y., Corsaro, B. & Hughes, P.R. (2001). Enhancement of Bacillus thuringiensis toxicity to Lepidopterous species with the enhancin from *Trichoplusia ni* granulovirus. *Biol. Con.* 20:153-159.

Hara, S., Tanada, Y. & Omi, E.M. (1976). Isolation and characterization of a synergistic enzyme from capsule of a granulosis virus of armyworm, *Pseudaletia-unipuncta*. *J. Invertebr. Pathol.* 27:115-124.

Harrison, R.L. (2009). Genomic sequence analysis of the Illinois strain of the *Agrotis ipsilon* multiple nucleopolyhedrovirus. *Virus Genes* 38:155-170.

Harrison, R.L. & Bonning, B.C. (2003). Comparative analysis of the genomes of *Rachiplusia ou* and *Autographa californica* multiple nucleopolyhedroviruses. *J. Gen. Virol.* 84: 1827–1842.

Harrison, R.L. & Lynn. D.E. (2007). Genomic sequence analysis of a nucleopolyhedrovirus isolated from the diamondback moth, *Plutella xylostella*. *Virus Genes* 35:857-873.

Harrison, R.L., Puttler, B., & Popham, J.J.R. (2008). Genomic sequence analysis of a fast-killing isolate of *Spodoptera frugiperda* multiple nucleopolyhedrovirus. *J. Gen. Virol.* 89:775-790.

Harrison, R.L. & Popham, H.J.R. (2008). Genomic sequence analysis of a granulovirus isolated from the Old World bollworm, *Helicoverpa armigera*. *Virus Genes* 36:565-581.

Harrison R.L., Sparks W.O. & Bonning B.C. (2010). *Autographa californica* multiple nucleopolyhedrovirus ODV-E56 envelope protein is required for oral infectivity and can be substituted functionally by *Rachiplusia ou* multiple nucleopolyhedrovirus ODV-E56. *J. Gen. Virol.* 91:1173–1182.

Häse, C. C. & R. A. Finkelstein. (1993). Bacterial extracellular zinc-containing metalloproteases. *Microbiol. Rev.* 57:823–837.

Hashimoto, Y., Corsaro, B.G. & Granados, R.R. (1991). Location and nucleotide sequence of the gene encoding the viral enhancing factor of the *Trichoplusia ni* granulosis virus. *J. Gen. Virol.* 72:2645-2651.

Hashimoto, Y., Hayakawa, T., Ueno, Y., Fujita, T., Sano, Y. & Matsumoto, T. (2000). Sequence analysis of the *Plutella xylostella* granulovirus genome. *Virology* 275: 358–372.

Hayakawa, R., Shimojo, E., Mori, M., Kaido, M., Furusawa, I., Miyata, S., Sano, Y., Matsumoto, T., Hashimoto, Y. & and Granados, R.R. (2000). Enhancement of baculovirus infection in *Spodoptera exigua* (Lepidoptera: Noctuidae) larvae with *Autographa californica* nucleopolyhedrovirus or *Nicotiana tabacum* engineered with a granulovirus *enhancin* gene. *Appl. Entomol. Zool.* 35:163–170.

Hegedus, D., Erlandson, M., Gillott, C. & Toprak, U. (2009). *New insights into peritrophic matrix synthesis, architecture, and function. Annu. Rev. Entomol.* 54:285–302.

Henikoff, S. & Henikoff, J.G. (1991). Automatic assembly of protein blocks for database searching. *Nucleic Acids Res.* 19:6565–6572.

Herniou, E.A., Olszewski, J.A., Cory, J.S. & O'Reilly, D.R. (2003). The genome sequence and evolution of Baculoviruses. *Annu. Rev. Entomol.* 48:211-234.

Herniou, E.A. & Jehle, J.A. (2007). Baculovirus phylogeny and evolution. *Curr. Drug Targets* 8:1043-1050.

Hoover, K., Humphries, M.A., Gendron, A.R., & Slavicek, J.M. (2010). Impact of viral enhancin genes on potency of *Lymantria dispar* nucleopolyhedrovirus in *L. dispar* following disruption of the peritrophic matrix. *J. Inverteb. Pathol.* 104:150-152.

Horton H.M. & Burand J.P. (1993). Saturable attachment sites for polyhedron-derived baculovirus on insect cells and evidence for entry via direct membrane fusion. *J. Virol.* 67:1860–1868.

Hayakawa, T., Shimojo, E., Mori, M., Kaido, M., Furusawa, I., Miyata, S., Sano, Y., Matsumoto, T., Hashimoto, Y & Granados, R.R. (2000). Enhancement of baculovirus infection in *Spodoptera exigua* (Lepidoptera: Noctuidae) larvae with *Autographa californica* nucleopolyhedrovirus or *Nicotiana tabacum* engineered with a granulovirus *enhancin* gene. *Appl. Entomol. Zool.* 35:163-170.

Hukuhara, T. & Zhu, Y. (1989). Enhancement of the in vitro infectivity of a nuclear polyhedrosis virus by a factor in the capsule of a granulosis virus. *J. Invertebr. Pathol.* 54, 71-78.

Hyink, O., Dellow, R.A., Olsen, M.J., Caradoc-Davies, K.M.B., Drake, K., Herniou, E.A., Cory,J.S., O'Reilly, D.R. & Ward, V.K. (2002). Whole genome analysis of the *Epiphyas postvittana* nucleopolyhedrovirus. *J. Gen. Virol.* 83:957–971.

IJkel, W.F.J., van Strien, E.A., Heldens, J.G.M., Broer, R., Zuidema, D., Goldbach, R.W., & Vlak, J.M. (1999). Sequence and organization of the *Spodoptera exigua* multicapsid nucleopolyhedrovirus genome. *J. Gen. Virol.* 80: 3289–3304.

Ikeda, M., Shikata, M., Shirata, N., Chaeychomsri, S. & Kobayashi, M. (2006). Gene organization and complete sequence of the *Hyphantria cunea* nucleopolyhedrovirus genome. *J. Gen. Virol.* 87:2549-2562.

Ivanova, N., SorokinA., Anderson, I., Galleron, N., Candelon, B., Kapatral, V., Bhattacharyya, A., Reznik, G., Mikhailova,N., Lapidus, A., Chu, L., Mazur, M., Goltsman, E., Larsen, N., D'Souza, M., Walunas, T., Grechkin, Y., Pusch, G., Haselkorn, R., Fonstein, M., Ehrlich, S.D., Overbeek, R. & Kyrpides. N. (2003). Genome sequence of *Bacillus cereus* and comparative analysis with *Bacillus anthracis*. *Nature* 423:87–91.

Jakubowska, A. K., Peters, S. A., Ziemnicka, J., Vlak, J. M. & van Oers, M. M. (2006). Genome sequence of an *enhancin* gene-rich nucleopolyhedrovirus (NPV) from *Agrotis segetum*: collinearity with *Spodoptera exigua* multiple NPV. *J. Gen. Virol.* 87: 537–551.

Jehle, J.A., Lange, M., Wang, H., Zhihong, H., Wang, Y. & Hauschild, R. (2006). Molecular identification and phylogenetic analysis of baculoviruses from Lepidoptera. *Virology* 346: 180-193.

Jiang, W., & Bond, J.S. 1992. Families of metalloendopeptidases and their relationships. *FEBS Lett.* 312:110–114.

Jongeneel, C. V., Bouvier, J. & Bairoch. A. (1989). A unique signature identifies a family of zinc-dependent metallopeptidases. *FEBS Lett.* 242:211–214.

Keddie B.A., Aponte G.W. & Volkman, L.E. (1989). *The pathway of infection of Autographa californica nuclear polyhedrosis virus in an insect host. Science* 243:1728–1730.

Kikhno, I., Gutierrez, S., Croizier, L., Croizier, G. & Ferber, M.L. (2002). *Characterization of pif, a gene required for the per os infectivity of Spodoptera littoralis nucleopolyhedrovirus. J. Gen. Virol.* 83:3013–3022.

Kozuma, K. & Hukuhara, T., (1994). Fusion characteristics of a nuclear polyhedrosis-virus in cultured-cells-time-course and effect of a synergistic factor and pH. *J. Invertebr. Pathol.* 63:63-67.

Kuzio, J., Pearson, M.N., Harwood, S.H., Funk, C.J., Evans, J.T., Slavicek, J.M. & Rohrmann, G.F. (1999). Sequence and analysis of the genome of a baculovirus pathogenic for *Lymantria dispar*. *Virology* 253:17–34.

Lange , M. & Jehle, J.A. (2003). The genome of the *Cryptophlebia leucotreta* granulovirus. *Virology* 317:220–236.

Lauzon, H.A.M., Jamieson, P.B., Krell, P.J.& Arif, B.M.(2005). Gene organization and sequencing of the *Choristoneura fumiferana* defective nucleopolyhedrovirus genome. *J. Gen. Virol.* 86:945–961.

Lauzon, H.A., Lucarotti, C,J,, Krell, P.J., Feng, Q., Retnakaran, A. & Arif, B.M. (2004). Sequence and organization of the *Neodiprion lecontei* nucleopolyhedrovirus genome. *J. Virol.* 78:7023–7035.

Lehane, M. J. (1997). Peritrophic matrix structure and function. *Annu. Rev. Entomol.* 42: 525-550.

Lepore, L. S., Roelvink, P.R. & Granados, R.R. (1996). Enhancin, the granulosis virus protein that facilitates nucleopolyhedrovirus (NPV) infections, is a metalloprotease. *J. Invertebr. Pathol.* 68:131-140.

Li, L., Donly, C., Li, Q.J., Willis, L.G., Keddie, B.A., Erlandson, M.A.,& Theilmann, D.A. (2002). Identification and genomic analysis of a second species of nucleopolyhedrovirus isolated from *Mamestra configurata*. *Virology* 297:226–244.

Li, Q., Donly, C., Li, L.,Willis, .LG., Theilmann, D.A. & Erlandson, M. (2002). Sequence and organization of the *Mamestra configurata* nucleopolyhedrovirus genome. *Virology* 294:106-121.

Li, X., Song, J., Jiang, T., Liang, C. & Chen, X. (2007). *The N-terminal hydrophobic sequence of Autographa californica* nucleopolyhedrovirus PIF-3 is essential for oral infection. *Arch Virol.* 152:1851–1858.

Luque, T., Finch, R., Crook, N., O'Reilly, D.R. & Winstanley, D. (2001). The complete sequence of the *Cydia pomonella* granulovirus genome. *J. Gen. Virol.* 82:2531–2547.

Ma, X-C., Shang, J-Y., Yang, Z-N., Bao, Y-Y., Xiao, Q. & Zhang, C-X. (2007). Genome sequence and organization of a nucleopolyhedrovirus that infects the tea looper caterpillar, *Ectropis oblique. Virology* 360:235-246.

Miller, L.K. (1997). Introduction to the Baculoviruses. In: *The Baculoviruses,* pp. 1-6. Edited by L.K. Miller. New York: Plenum.

Moscardi, F. (1999). Assessment of the applications of baculoviruses for control of Lepidoptera. *Annu. Rev. Entomol.* 44:257-289.

Murphy, G. J. P., Murphy, G. & Reynolds, J.J. (1991). The origin of matrix metalloproteinases and their familial relationships. *FEBS Lett.* 289:4–7.

Nai, Y-S., Wu, C-Y., Wang, T-C., Chen, Y-R., Lau, W-H., Lo, C-F., Tsai, M-F. & Wang, C-H. (2010). Genomic sequencing and analysi of *Lymantria xylina* mutitiple nucleopolyedtovirus. *BMN Genomics* 11:116.

Nakai, M., Goto, C., Kang, W., Shikata, M., Luque, T. & Kunimi, Y. (2003). Genome sequence and organization of a nucleopolyhedrovirus isolated from the smaller tea tortrix, *Adoxophyes honmai. Virology* 316:171–183.

Ohkawa, T., Washburn,J.O., Sitapara, R., Sid, E. & Volkman, L.E. (2005). *Specific binding of Autographa californica* M nucleopolyhedrovirus occlusion-derived virus to midgut cells of Heliothis virescens larvae is mediated by products of pif genes Ac119 and Ac022 but not by Ac115. *J. Virol.* 79:15258–15264.

Ohkawa, T., Volkman, L.E. & Welch, M.D. (2010). *Actin-based motility drives baculovirus transit to the nucleus and cell surface. J. Cell Biol.* 190:187–95.

Oliveira, J.V.C., Wolff, J.L.C., Garcia-Maruniak, A., Rebeiro, B.M., de Castro, M.E.B., de Souza, M.L., Moscardi, F., Maruniak, J.E. & Zanotto, P.M.A. (2006). Genome of the most widely used viral biopesticide: *Anticarsia gemmatalis* multiple nucleopolyhedrovirus. *J. Gen. Virol.* 87:3233-3250.

Pang, Y., Yu, J.X., Wang, L.H., Hu, X.H., Bao, W.D., Li, G., Chen, C., Han, H., Hu, S.N. & Yang, H.M. (2001). Sequence analysis of the *Spodoptera litura* multicapsid nucleopolyhedrovirus genome. *Virology* 287:391–404.

Parkhill, J., Wren, B.W., Thomson, N.R., Titball, R.W., Holden, M.T., Prentice, M.B., Sebaihia, M., James, K.D., Churcher, C., Mungall, K.L., Baker, S., Basham, D., Bentley, S.D., Brooks, K., Cerdeno-Tarraga, A.M., Chillingworth, T., Cronin, A.,Davies, R.M., Davis, P., Dougan, G., Feltwell, T., Hamlin, N., Holroyd, S., Jagels, K., Karlyshev, A.V., Leather, S., Moule, S., Oyston, P.C., Quail, M., Rutherford, K., Simmonds, M., Skelton, J., Stevens, K., Whitehead, S. & Barrell, B.G. (2001). Genome sequence of *Yersinia pestis,* the causative agent of plague. *Nature* 413:523–527.

Peng, K., van Oers, M.M., Hu, Z., van Lent, J.W. & Vlak, J.M. (2010). Baculovirus per os infectivity factors form a complex on the surface of occlusion-derived virus. *J. Virol.* 84:9497–9504.

Peng, J., Zhong, J. & Granados, R.R. (1999). A baculovirus enhancin alters the permeability of a mucosal midgut peritrophic matrix from lepidopteran larvae. *J. Insect Physiol.* 45:159–166.

Popham, H. J. R., Bischoff, D.S. & Slavicek, J.M. (2001). Both *Lymantria dispar* nucleopolyhedrovirus *enhancin* genes contribute to viral potency. *J. Virol.* 75:8639-8648.

Pritchett, D. W., Young, S. Y. & Yearian, W. C. (1984). Some factors involved in the dissolution of *Autographa californica* nuclear polyhedrosis virus polyhedra by digestive fluids of *Trichoplusia ni* larvae. *J. Invertebr. Pathol.* 43:160–168.

Qianjun, Li., Li, L., Moore, K., Donly, C., Thielmann, D.A. & Erlandson, M. (2003). Characterization of *Mamestra configurata* nucleopolyhedrovirus enhancin and its functional analysis via expression in an *Authographa californica* M nucleopolyhedrovirus recombinant. *J. Gen. Virol.* 84:123-132.

Read, T. D., Peterson, S.N., Tourasse, N., Baillie, L.W., Paulsen, I.T., Nelson, K.E., Tettelin, H., Fouts, D.E., Eisen, J.A., Gill, S.R., Holtzapple, E.K., Okstad, O.A., Helgason, E., Rilstone, J., Wu, M., Kolonay, J.F., Beanan, M.J., Dodson, R.J., Brinkac, L.M., Gwinn, M., DeBoy, R.T., Madpu, R., Daugherty, S.C., Durkin, A.S., Haft, D.H., Nelson, W.C., Peterson, J.D., Pop, M., Khouri, H.M., Radune, D., Benton, J.L., Mahamoud, Y., Jiang, L., Hance, I.R., Weidman, J.F., Berry, K.J., Plaut, R.D., Wolf, A.M., Watkins, K.L., Nierman, W.C., Hazen, A., Cline, R., Redmond, C., Thwaite, J.E., White, O., Salzberg, S.L., Thomason, B., Friedlande, A.M., Koehler, T.M., Hanna, P.C., Kolsto, A.B. & Fraser, C.M. (2003). The genome sequence of *Bacillus anthracis* Ames and comparison to closely related bacteria. *Nature* 423:81–86.

Rohrmann, G.F. (2011). Baculovirus Molecular Biology: Second Edition. Bethesda (MD): National Center for Biotechnology Information (US); Bookshelf ID: NBK49500 (http://www.ncbi.nlm.nih.gov/books/NBK49500/).

Slavicek, J.M. & Popham, J.R.. (2005). The *Lymantria dispar* nucleopolyhedrovirus enhancins are components of occlusion-derived virus. *J. Virol.* 79: 10578-10588.

Szewczyk, B., Hoyos-Carvajal, L., Paluszek, M., Skrzecz, I. & Souza, M.L. (2006). Baculovirus - re-emerging biopesticides. *Biotechnol. Adv.* 24:143-160.

Szewczyk, B., Rabalski, L., Krol, E., Sihler, W., & Souza, M.L. (2009). Baculovirus biopesticides - a safe alternative to chemical protection of plants. *J. Biopestic.* 2:209-216.

Tanada, Y. (1959). Synergism between two viruses of the armyworm, *Pseudaletia unipuncta* (Hawworth) (Lepidoptera, Noctuidae). *J. Insect Pathol.* 1:215-231.

Tanada, Y. (1985). A synopsis of studies on the synergistic property of an insect baculovirus: A tribute to Edward A. Steinhaus. *J. Invertebr. Pathol.* 45:125-138.

Tanada, Y., Hess, R.T. & Omi, E.M. (1975). Invasion of a nuclear polyhedrosis virus in midgut of the armyworm, *Pseudaletia unipuncta,* and the enhancement of a synergistic enzyme. *J. Invertebr. Pathol.* 26:99-104.

Tanada, Y., Himeno, M. & Omi, E.M. (1973). Isolation of a factor, from the capsule of a granulosis virus, synergistic for a nuclear-polyhedrosis virus of the armyworm. *J. Invertebr. Pathol.* 21:31-40.

Tanada, Y., Inoue, H., Hess, R.T., & Omi, E.M. (1980). Site of action of a synergistic factor of a granulosis virus of the armyworm, *Pseudaletia unipuncta. J. Invertebr. Pathol.* 34: 249-255.

Tang, X-D., Xiao, Q., Ma, X-C., Zhu, Z-R. & Zhang, C-X. (2009). Morphology and genome of *Euproctis pseudoconspersa* nucleopolyhedrovirus. *Virus Genes* 38:495-506.

Terra, W. R. (2001). The origin and functions of the insect peritrophic membrane and peritrophic gel. *Arch. Insect Biochem. Physiol.* 47:47-61.

Uchima, K., Harvey, J.P., Omi, E.M. & Tanada, Y. (1988). Binding sites on the midgut cell membrane for the synergistic factor of a granulosis virus of the armyworm (*Pseudaletia unipuncta*). *Insect Biochem.* 18:645-650.

van Oers, M. M., Abma-Henkens, M. H., Herniou, E. A., de Groot, J. C., Peters, S. & Vlak, J. M. (2005). Genome sequence of *Chrysodeixis chalcites* nucleopolyhedrovirus, a baculovirus with two DNA photolyase genes. *J. Gen. Virol.* 86:2069–2080.

Volkman, L. E. (2007). Baculovirus infectivity and the actin cytoskeleton. *Curr. Drug Targets* 8:1075–1083.

Wang, P. & Granados, R. R. (1997). An intestinal mucin is the target substrate for a baculovirus enhancin. *Proc. Natl. Acad. Sci. USA* 94:6977-6982.

Wang, P. & Granados, R. R. (2000). Calcofluor disrupts the midgut defense system in insects. *Insect Biochem. Mol. Biol.* 30:135-143.

Wang,Y., Choi,J.Y., Roh,J.Y., Woo,S.D., Jin,B.R. & Je,Y.H. (2008). Molecular and phylogenetic characterization of *Spodoptera litura* Granulovirus. *J. Microbiol.* 46:704-708.

Willis, L.G., Siepp, R., Stewart, T.M., Erlandson, M.A. & Theilmann, D.A. (2005). Sequence analysis of the complete genome of *Trichoplusia ni* single nucleopolyhedrovirus and the identification of a baculoviral photolyase gene. *Virology* 338:209–226.

Wormleaton, S., Kuzio, J. & Winstanley, D. (2003). The complete sequence of the *Adoxophyes orana* granulovirus genome. *Virology* 311:350-365.

Xiao, H. & Yipeng, Q. (2007). Genome sequence of *Leucania seperata* nucleopolyhedrovirus. *Virus Genes* 35:845-856.

Yamamoto, T. & Tanada, Y. (1978). Protein components of two strains of granulosis virus of the armyworm, *Pseudaletia unipuncta* (Lepidoptera: Noctuidae). *J. Invert. Pathol.* 32:158-170.

Yamamoto, T. & Tanada, Y. (1980). Physiochemcial properties and location of capsule components , in particular the synergistic factor, in the occlusion body of a granulosis virus of the armyworm, *Pseudaletia unipuncta. Virology* 107:434-440.

Zhu, S-Y., Yi, J-P., Shen, W-D., Wang, L-Q., He, H-G., Wang, Y., Li, B & Wang, W-B. (2009). Genomic sequence, organization and characteristics of a new nucleopolyhedrovirus isolated from *Clanis bilineata* larva. *BMC Genomics* 10:91.